Patrik Lindenfors

For Whose Benefit?

The Biological and Cultural Evolution of Human Cooperation

Springer

Patrik Lindenfors
Centre for the Study of Cultural Evolution
Stockholm University
Stockholm
Sweden

ISBN 978-3-319-50873-3 ISBN 978-3-319-50874-0 (eBook)
DOI 10.1007/978-3-319-50874-0

Library of Congress Control Number: 2016960026

Translation from the Swedish language edition: *Samarbete* by Patrik Lindenfors, © Patrik Lindenfors 2011.

Printed on acid-free paper

This Springer imprint is published by Springer Nature
The registered company is Springer International Publishing AG
The registered company address is: Gewerbestrasse 11, 6330 Cham, Switzerland

For Anna

Contents

The Human Puzzle

Abstract We are made of cooperating parts, both physically and psychologically. But what exactly is cooperation? Is cooperation on all the levels that make up humans and societies a similar process? The main idea is introduced, that cooperation resulting in emergent larger structures has common characteristics:

1. Similarity of cooperating parts.
   (a) In fraternal cooperation, cooperating parts are on a fundamental level alike.
   (b) In egalitarian cooperation, cooperating parts are not alike.
2. On each level, there exist control mechanisms that handle constituent parts that, having lost their independence, start acting egoistically.
   (a) In fraternal cooperation, such internal control mechanisms exist for culling or correcting overly 'selfish' genes, cancerous cells, criminal individuals, and bad ideas.
   (b) In egalitarian cooperation, mutual dependence functions as a similar control mechanism—breaking cooperation entails an immediate loss.
3. On every level, there is a surplus of redundant parts, resulting in robustness and resilience.
4. The constituent cooperating parts do not always reap immediate benefits from the cooperation, but sometimes the cooperation instead benefits some other, underlying unit which is the one that really benefits from the cooperation. *All* cooperation benefits the copying of the replicators—genes or memes.
5. Emergent properties, where the whole is more than just the sum of its constituent parts, are a result of specialization in the constituent parts.
6. Cooperation creates order from disorder by repeating the same procedure over and over.

P. Lindenfors, *For Whose Benefit?*, DOI 10.1007/978-3-319-50874-0_1

A few summers ago I helped my dad lay a new barn floor. After finishing a particularly tricky floorboard he put down his hammer and looked at me. 'You're *so* much easier to cooperate with than your brother…' For a second I thought he was going to pay me a compliment. '…because you do what I tell you.'

Although a bit disappointed by the lack of positive feedback I knew exactly what he was referring to. My brother, when he was still alive, always had his own ideas about how things were to be done. Even the tiniest design issue had to be discussed, sometimes endlessly. My brother cooperated well with no one, least of all himself.

I, on the other hand, just went along with what my father asked of me in these building situations. I have no particular opinion on how to lay barn floors, or how best to fix a broken door, or how to repair the water pump. If my dad wanted it some specific way, that's the way he would get it. What my father interpreted as good cooperation was merely calculated disinterest on my part. I just wanted the work over and done with so I could go back to reading my books.

But was my father correct in terming my unquestioning compliance cooperation? On one hand obviously yes, we were doing something together—co-operating—thus achieving something together that we could not have achieved singly. On the other hand obviously no; submitting to the will of another is not what we mean by cooperation. So why was I working when I really wanted to read?

If instead I had been locked inside a prison and exploited as forced labor, then the exact same floor-laying would not have been cooperation at all, but punishment. Had I been enslaved, the same work would have counted as parasitism. Yet again, if I had been laying the floor as paid labor, I would have been participating in a market transaction—my labor in exchange for money. This is definitely a form of cooperation, through the mutual exchange of commodities.

Perhaps my motivation was selfish—I was working to secure my inheritance. Unless, of course, I did it for my children, in which case it could be viewed as an investment in my genetic interests, in my legacy.

Or perhaps I was just a victim of cultural circumstances, of the notion that *of course* you help your aging parents. What would other people think of me if I didn't? Would they in turn refrain from helping me the day that I would need help myself?

Clearly, it is not possible to simply look at our behavior to determine if we were cooperating or if we were examples of something else, such as parasitism, coercion, punishment, egoism, spite or cultural tradition. To understand cooperation we must understand motivations.

But the central problem of understanding cooperation is this: How can we understand situations where one individual does something he doesn't want to do, has no interest in doing, or has to pay to do, for the benefit of another individual? *Under what circumstances will an individual take a cost for another's benefit*?

As we saw, the simplest explanation may be pure self-interest: I may be working to protect my inheritance. This kind of cooperation is easy to understand since there is no 'real' cost involved in the end as the cost is only temporary—an investment.

On the other hand, a biologist may be keen to point out that my father and I are kin, and that our mutual giving and taking of favors therefore occurs within a well understood biological context. For reasons having to do with evolution, sharing genes will lead to an expectation that we will be willing to take costs for one another. This is by far the most common form of cooperation seen in nature.

But we should not discount the fact that my parents took care of me, fed me, protected me and paid for me when I was a helpless child and all through adolescence. Part of what I was doing that day, building the barn floor, was reciprocation. Since my father was so old that he could not carry out all repairs on his own (he was over 90!) it wasn't more than fair that I helped a bit (although to be really honest my sister does most of the helping). Such reciprocity is the second most common reason for cooperation.

Common interests, relatedness and reciprocity explain almost all cooperation we find in the natural world. But in humans there is more, much more. Think, for example, of what would have happened had I not helped my father and people heard of this. I would get a reputation that I was a selfish bastard, something that could harm my future interactions with other people, people who know me or my family. This is termed *indirect reciprocity*, that two persons' interactions not only influence future interactions between those two individuals, but that our reputation has repercussions outside the immediate situation.

There are more ways than these to understanding cooperation in humans. Most notably these concern cultural group selection and spatial selection; that we can sacrifice ourselves for groups we belong to, such as nations, religions or ideologies, and that we tend to cooperate with those in our proximity, or move closer to those we cooperate well with.

All explanations, however, are aimed at understanding the same crucial dilemma: the conditions for when one individual will absorb a cost for another. Thus *all* proposed explanations come down to answering the question in the title: For whose benefit? Or, if you prefer Latin: *Cui bono*?

In cases in nature where there seems to exist no short- or long-term benefit to the self-sacrificing individual—the individual taking the cost—the answer can instead come down to benefits for lower level units, what Richard Dawkins has termed *replicators*: genes or 'memes'.

While genes are rather well understood biological units of inheritance, memes are less well accepted, to the point that the scientific *Journal of Memetics* has shut down due to a lack of progress. Memes are undefined units of cultural inheritance: 'that which is copied' when we learn something. Since this unit is so disputed it has to be discussed with great care, something I will attempt to do in the latter half of the book. Most researchers currently shy away from using the term 'meme' while continuing to analyze the spread of cultural knowledge *as though* there really existed particles of culture that spread across populations.

But cooperation is not only about cooperating individuals; cooperation is one of the fundamental principles of life itself; a process that is present on every level of organization in your body, from your genes up to your conscious 'self'.

The Breakdown of Self

When you look at yourself in the mirror, how many do you see? Most of us see *one* individual—*one* self. But a description which is just as true is that we all are three-dimensional mosaics of cooperating cells—a human jigsaw puzzle. This is knowledge that we acquire some time in school without it impacting our picture of ourselves the least. *Of course* the great unity that is our physical selves consists of smaller parts. So what?

But cells are not the end of it. Each and every cell in your body is in turn built by cooperating parts—organelles (cellular 'organs'). Some of these even have their own separate DNA. Moreover, each cell is governed by its own genes, which cooperate with each other to build, run and repair themselves, the cells and the body to which they belong.

Speaking of genes, fact is that most of the DNA that you carry around isn't even your own, but belongs to bacteria that live in your guts, bacteria that cooperate with you in digesting food. So who is that single unit you spot in the mirror—the one worthy of being called 'I'? Can all bacteria be removed—the ones that help you digest your food, for example—or are they also part of the unit that is *you*?

If you believe that the answer is in some way self-evident—that *of course* you can't count the bacteria as part of your *I*, then contemplate, for a moment, how many other parts that can be removed before your identity is lost.

What If you chopped off your arm? Would there then be two *I*, or is the crucial *I* securely locked inside your skull—in the brain—free of the arm? So says the brain. The arm can't go on living without the rest of the body and your consciousness surely seems somehow lodged inside your head, so the answer should be that there is one *I* that can do just as well without the arm. But how many body parts can you lose before it starts making a crucial difference?

It turns out that you can do without large parts of the brain too, without it having too large an impact on your behavior. Where is then this *I* if the whole brain isn't even needed?

The problem is that there is no '*I*-spot' inside the head, no special place, no special nerve cell, where someone sits and is *you*. Instead, your brain is built out of about 86 billion nerve cells (Azavedo et al. 2009), where each nerve cell is in contact with roughly 10,000 other nerve cells. These cells and connections receive and handle light impulses from the mirror, compare the composite image with different memories ('Don't I look really great today?') and present emotional and intellectual reactions ('No').

Hold on. Present to whom? There must be someone to present these reactions to, but if there is no *I*-spot then there cannot exist any such viewer. A better way of putting it must therefore be that the emotional and intellectual reactions *are* you, or at least a part of you, where another part of you uses this incoming information to construct an appropriate response.

So both your physical and mental *I* seem to be built out of parts. Is that how we should understand the brain and the body, as parts of you that you can do without, but that nonetheless are part of that 'you'?

But if the *I* consists of parts, then who are *you*?

Such issues are a bit confusing to get a handle on, but only because we are trying to understand ourselves through introspection. That parts can cooperate to build up a whole, where the composite unit is larger than its constituent parts, is nothing strange if we consider for example the fact that several people kicking a ball together make up a team.

This is what we normally mean by cooperation: single individuals that together carry out some task. Together we can for example make up a family, inhabitants of a country, car drivers, or employees at Starbucks. All these groups of people are separate units with their own organization and explicit or implicit rules for collaboration, units where we adapt specific positions with different roles and expectations.

So cooperation obviously exists on many levels. Is there any unifying principle? If you look up 'cooperation' in the Merrian-Webster English thesaurus you find the following:

> 1: to act or work with another or others: act together or in compliance
> 2: to associate with another or others for mutual benefit

Note that according to this definition, there is no conscious action needed for cooperation, but mutual benefit is necessary. If there is only benefit to one party it is instead parasitism or coercion. By this definition, then, it is ok to look at the cells that make up your body as cooperating units that are doing something together since they gain mutual benefit from doing so, even if they are not consciously choosing to cooperate.

It is according to this definition that this book will look into cooperation:

> Cooperation is collective functioning of some kind of units for the benefit of themselves and/or their component parts.

Cooperation is a form of collective action, where the consequences of several units functioning together results in something that wouldn't have happened if there had been no collective action. As we are looking at *evolutionary* problems, costs and benefits are measured by the number of copies of the replicators (genes or 'memes') relative to non-collectively acting units.

Thus, an important point is that the collectively functioning units need not be the same units that derive benefit from the collaboration. Cells can for example suffer programmed cell death for the benefit of the whole organism, since every cell in the organism has the same genes—cells die so that their genes can multiply better. The problem with understanding how cooperation has evolved and how it can work is often in understanding which units benefit from the collective action. Sometimes the answer is self-evident, at other times one needs a rather deep understanding of the theory of evolution. And, as we shall see, some forms of cooperation are still not well understood.

I will start with the description of cooperation *within* individuals: how we are built from the inside out of conglomerate upon conglomerate of interlocking cooperating parts. Later, I will describe cooperation *between* individuals and how it sometimes can be understood as a consequence of the cooperation within. I also want to highlight at what point cooperation between many smaller units results in a new, larger unit, with its own characteristics not predictable from the component parts, as when cells make up individuals, nerve cells make up subjective experience, individuals make up a society, or many ideas (or memes, if you like) make up cultural systems.

Emergent units across many levels seem to have comparable organization, independent of whether one looks at the cooperation that results in the genome, the eukaryote cell, the body, society, or our common knowledge. On almost every level where parts make up a unit larger than the sum of the constituent parts at least six observations of similarity can be made.

American biologist David Queller has made a useful distinction between 'fraternal' and 'egalitarian' cooperation. Egalitarian cooperation is between unlike components, where two different kinds of entities come together bringing different capacities that work well in combination (our bacteria and ourselves, for example). Fraternal cooperation is between like components where like kinds of entities work together for mutual benefit (our body-cells, for example) (Queller 1997). Differences exist between these two types of cooperation, as indicated in the list below.

1. Similarity of cooperating parts.
   (a) In fraternal cooperation, the cooperating parts are on a fundamental level alike. Genes cooperate with genes, cells cooperate with cells, nerve cells with nerve cells, individuals with individuals, and memes with memes.
   (b) In egalitarian cooperation, the parts are not alike. Different kinds of organelles cooperate in a cell, different species cooperate in symbiotic relationships, and different professions cooperate in a society.
2. On each level, there exist control mechanisms that handle constituent parts that, having lost their independence, start acting egoistically, i.e. that start functioning in a way furthering their own benefit at the detriment of the whole.
   (a) In fraternal cooperation, such internal control mechanisms exist for culling or correcting overly 'selfish' genes, cancerous cells, criminal individuals, and bad ideas.
   (b) In egalitarian cooperation, mutual dependence functions as a similar control mechanism—breaking cooperation entails an immediate loss of collective functioning.
3. On every level, there is a surplus of redundant parts, something resulting in robustness and resilience—many parts can break without causing the whole to fail.

4. The constituent cooperating parts do not always reap immediate benefits from the cooperation, but sometimes the cooperation instead benefits some other, underlying unit which is the one that really benefits from the cooperation. This can result in cooperation that at first view can seem self-sacrificial. To solve such puzzles, it is necessary to answer the question '*For whose benefit?*' As we shall see, *all* cooperation benefits the copying of the replicators—genes or memes.
5. Emergent properties, where the whole is more than just the sum of its constituent parts, are partly a result of specialization in the constituent parts.
6. Cooperation creates order from disorder by repeating the same procedure over and over. Genes, cells and memes are copied; individuals reproduce. Genes are structured in DNA, cells are structured in organs, neurons are structured in brain components, humans are structured in groups (e.g. ethnic or professional groups), memes are structured into different fields of knowing or cultural constructs.

Cooperation and Life

With the more encompassing definition of cooperation—that Cooperation is collective functioning of some kind of units for the benefit of themselves and/or their component parts—there is much that counts as cooperation. Too much?

Is it, for example, correct to state that nature's constants cooperate to make up the universe? The British astrophysicist Martin Rees has argued that there are six constants of nature—six cosmic numbers—that determine the form, size and make-up of the universe (Rees 1999). Had one of these constants been only slightly different, then life as we know it wouldn't have been able to exist. Humankind would not have existed and thus no one would have been able to reflect on the constants of nature. Since these constants of nature determine the rules for our existence, is it therefore correct to claim that the constants of nature cooperate to build the universe?

No. Ask yourself what benefit these constants would derive from making up a universe. None whatsoever. Consequently, according to our definition, because there exists no benefit, it is not cooperation.

So what about chemical bonds between atoms? Take water: two oxygen atoms have six electrons in their outermost electron layer and hydrogen atoms only have one, oxygen needs eight to fill its outermost layer, while each hydrogen atom needs two. When two hydrogen atoms fuse with one oxygen atom, the outermost electron layer of the oxygen atom is filled with eight electrons (6 + 1 + 1) because the electrons are shared between the oxygen and the hydrogen atoms. Similarly, each hydrogen atom gets two electrons in its outermost layer. A stable molecule is formed—water.

So what kind of benefit do hydrogen and oxygen atoms have from fusing into water? Possibly to achieve a full outermost electron layer—using a very wide definition of 'benefit' you might count this as cooperation.

But if this is enough to count as 'benefit', then all chemical reactions, atomic and sub-atomic interactions must count as cooperation. The boundary has to be specified more exclusively or I would have to write a very thick book indeed. I will here draw the boundary at life. But what exactly is life?

Life

The British author Douglas Adams summarized the problem with trying to understand life in a characteristically drastic way. 'If you try and take a cat apart to see how it works, the first thing you have on your hands is a nonworking cat.' (Dawkins 2001).

Yet on another level the question may sound strange. If some person you were talking to were to fall dead to the ground you would most definitely note the difference. We all know what life is, at least intuitively. But we don't know what exactly becomes different when someone dies, what disappears.

The speculations around this through the millennia have been many. Most of these have involved some form of life force, an *élan vital*, which is lost at the moment of death. Life, according to this view, is matter plus 'something more'.

To believe that life besides matter also consists of 'something more' opens a door for all kinds of speculation. One historical suggestion has been that this 'something' can communicate with us from 'the other side', or be eternally tormented.

The scientific view is clear. Life is not matter and 'something more'. Life is just matter—a particularly interesting set of chemical reactions—and the border between the living and the non-living is pretty blurry at that. If you look up 'life' in an encyclopedia you therefore normally find not a definition but a list of characteristics that are associated with life.

- Metabolism.
- Reproduction.
- The creation of order from non-order (entropy reduction).
- The ability to react to the environment.
- Inheritance that is transmitted through DNA or RNA.[1]
- Evolution.

[1]DNA is the spiral formed molecule that transmits the genetic inheritance. DNA has four base pairs that serve as letters in the genetic alphabet: adenine (A), tymine (T), guanine (G) and cytosine (C). RNA is (most often) single stranded DNA with uracile instead of thymine. All life has inheritance that is transmitted through DNA, except some viruses and all viroids that instead have inheritance transmitted through RNA. In cells, the instructions are translated from DNA via different forms of RNA to proteins.

The most common examples of border states between the living and the dead are viruses and viroids (viruses without 'shells'). None of these have capacity for metabolism or reproduction on their own and can therefore not create order or register anything about the environment. Viruses and viroids instead hi-jack cells of other organisms by getting themselves incorporated into the host cells' DNA and reprogramming the cell metabolism to make more copies of themselves—this is what happens when, for example, you catch a cold.

Concerning metabolism, reproduction and the ability to register information about the environment, viruses and viroids are not life. But they do have inheritance that is transmitted through DNA or RNA, and they do evolve. Therefore, viruses and viroids are often considered a borderline case, being neither living nor non-living. (An easy way to communicate with the non-living is thus to come down with a good cold.)

Perhaps the molecular configuration that amounts to COPY ME is the best criterion for life. As soon as such a copying instruction emerges, an almost unstoppable process starts—evolution. The chemical reaction that results in self-replication may thus be the most important and most basal of all the characteristics of life.

Viruses and viroids have inheritance that makes use of such copying instructions, although they cannot carry out the instruction without 'borrowing' the molecular machinery of other cells. They could therefore be considered life. However, with such a definition of life, computer viruses too have to be considered living organisms—they also have inheritance of instructions for making copies of itself, but need outside machinery to carry out the copying.

There is no single definition of life, it is a simple as that. I thus start with the unit that provides the copying instruction: genes—do genes cooperate? Hasn't Richard Dawkins stressed that they are selfish (Dawkins 1976)?

References

Azevedo, F. A. C., Carvalho, L. R. B., Grinberg, L. T., Farfel, J. M., Ferretti, R. E. L., Leite, R. E. P., et al. (2009). Equal numbers of neuronal and nonneuronal cells make the human brain an isometrically scaled-up primate brain. *Journal of Comparative Neurology, 513*, 532–541.

Dawkins, R. (1976). *The selfish gene*. Oxford: Oxford University Press.

Dawkins, R. (2001). Eulogy for Douglas Adams. September 17, 2001. http://www.edge.org/documents/adams_index.html

Queller, D. C. (1997). Cooperators since life began. *The Quarterly Review of Biology, 72*, 184–188.

Rees, M. (1999). *Just six numbers: The deep forces that shape the universe*. London: Weidenfeld & Nicolson.

Your Physical Self

Abstract The cooperation that results in our physical bodies is described; between genes, between prokaryotes, within eukaryotes, between eukaryotes (giving rise to multicellularity), and between the bacteria in our guts and ourselves. Our physical selves consists of conglomerates of layer upon layer of cooperating parts.

Genes

Everything biological is made of parts. But a single atom is not sufficient to give the instruction COPY ME (if that were enough, matter would replicate all around us—it doesn't). Such an instruction demands many atoms together, organized in special molecules: DNA or RNA. What these molecules do is to reorganize matter into copies of themselves. This is the most basic chemical reaction of life.

Think about the instruction COPY ME in a computer program. If you spell this instruction the least bit wrong, the computer will grind to a halt. Computers are stupid in that way and so is DNA. All the letters have to be there, united as the letters of a properly spelt word. If a piece of the code, such as OP ME, is on its own, it will soon disappear, uncopied.

Life based on the copying instructions in DNA works in exactly this way; the entire copying instruction is needed for the process to work. *Each* letter must co-exist with the others—they must function together. All instructions do not necessarily have to co-exist on the same molecule, however, as it has been shown that mixtures of RNA fragments spontaneously form cooperative catalytic cycles and networks (Vaidya et al. 2012).

Nevertheless, as an effect of this collective functioning, a unit emerges which is larger than the sum of its constituent parts, a *self-replicating molecule*; a unit that achieves something its constituent parts couldn't. The 'benefit' that the molecule reaps through the collective functioning of its parts is that it is replicated; more

P. Lindenfors, *For Whose Benefit?*, DOI 10.1007/978-3-319-50874-0_2

copies of the whole molecule are created. This is what is implied by 'benefit' in evolutionary terms.

Through being copied in greater frequency if things go right, and in lower frequency if they don't, costs and benefits now exist that can be counted and compared. 'Good' cooperation is cooperation that results in more copies of the self-replicating molecule, and 'bad' cooperation is cooperation that results in fewer copies of the self-replicating molecule. This is a criterion of benefit that will remain with us throughout the book.

But why attach value terms such as 'good' or 'bad' to differences in frequencies? This can of course be disputed, but genes with instructions resulting in more copies ('good') will replace genes with instructions resulting in fewer copies ('bad')—such competition is natural selection. What I term 'good' is winning this contest.

Self-copying is life's most basic instruction, the very basis of evolution. With genes, we have arrived at something we can term a *replicator*—a unit that copies itself. Richard Dawkins has defined replicator more stringently as 'any entity in the universe which interacts with its world, including other replicators, in such a way that copies of itself are made' (Dawkins 1978).

This replicator consists of smaller collectively functioning units that do not become copied when they are on their own. The unity of the constituent parts thus results in something more than just the sum of the parts.

What the replicator results in is nothing less than a chemical mechanism that orders matter through the biochemical manufacturing of more copies of itself, through the assembly of separate parts. From being a scrambled mess of parts, order is achieved through relentless repetitive copying. Thus, from the very beginning, life seems to defy the fundamental physical process that results in increased entropy, that results in a larger degree of disorder (of course, life doesn't really defy thermodynamics—energy is supplied from the outside, ultimately from the sun). This ordering capacity is one of the signs of life.

Organisms order matter by utilizing energy from their environment. Animals and mushrooms absorb order from other life forms when metabolizing or decomposing. The waste products from these processes are less ordered than what went in; this absorption of order drives the process. Plants utilize energy through transforming energy present in sunrays to energy stored in chemical bonds through the process known as photosynthesis. In this process, one of the waste products is heat—less ordered energy than what went in. Through their energy transformations animals and plants can build structures—leaves, stems, flowers, body parts, and in this way order matter on an even higher level than simply making copies of their DNA.

Some researchers believe that the initial copying instruction was given by a piece of randomly assembled RNA. Others hold that the first RNA instruction built on earlier types of self-replicating elements. At present, we simply do not know. What we do know is that the components necessary for life to emerge can emerge spontaneously, but we do not know how these components first assembled into a self-replicating molecule. Research on this topic is on-going.

With molecular copying there is immediately a possibility that the copying goes wrong in different ways. Sometimes the copy becomes so riddled with errors that that the copying ends (analogous to a misspelling: CPRK ME). However, the copying error can also make no real difference, but occur in such a way that the copying can keep working (analogous to a superfluous letter COPY ME K).

Had copying been carried out without error from the beginning, life would have stopped at the initial stage and no evolution taken place. Earth would have filled up with the same self-replicating molecule until all building blocks had been utilized and that would have been the end of it. Without errors there is no possibility of improvement. For evolution to occur, copying can't be perfect all the time, but can equally well not be erroneous all the time. The right amount of errors is a crucial part of the evolution of life: mutations.

A not uncommon error in the genome is something termed gene duplication, that one gene is copied twice (COPY ME|COPY ME). This means that mutations that previously had been directly lethal for copying can now take place in one gene, while the other remains intact and carries out the instruction. This at once opens up for new instructions (for example, COPY ME|CORRECT ME). In this way, new adaptations can creep into the genome.

Already here one central component of cooperation emerges: specialization. Even though genes are the same physical phenomena (and thus an example of fraternal cooperation—cooperation between like parts), different genes carry different information and thus have different functions and consequences. This is one of the reasons that the sum of many different genes results in something so much larger than just the sum of its parts.

Some genes in our bodies carry instructions to check that the copying process of the DNA has been carried out correctly. Others carry instructions for the protection of the genome through a secure cell wall. Yet others carry instructions for acquiring energy, or getting rid of excrement, and so on.

Note the emphasis on instructions—genes carry *information*. Their chemical properties have consequences above and beyond just affecting themselves. In animals, genes for example carry information for traits such as color, tooth size and claw design, but also behavioral traits such as aggressiveness, territory size and mating displays, and even traits outside the body, such as the design of beaver dams, bird nests and termite mounds.

The proper manner to regard genes is thus as *information carriers*, not just interesting chemical compounds that happen to copy themselves. The influential American biologist George C. Williams stressed this forcefully early on, pointing out that like poems and food recipes, the importance of genes lies in the information transferred, not the exact medium that carries the information. Just as it isn't important how much a poem or recipe weighs, if it is written down on paper, memorized, or hacked into stone, likewise the importance of genetic replicators are that they are information carriers, not that they are strings of nucleotides (Williams 1992).

Mutation speed is central already from life's very beginning. If random change happens too often, the copying system breaks down and the line of inheritance stops. If random change happens too seldom, no improvements occur and the line of inheritance is overtaken by others having greater variation.

In later forms of life, in more complicated organisms such as humans, mutation speed is different in different parts of the genome. Parts of the human genome are so critical to our survival and well-being that there is no room for error at all. But of course, sometimes errors happen anyway, and things go really bad. The copying stops: death.

One example from the human genome is the disease retinoblasma, a form of eye cancer in children. In somewhat less than half of the cases, a mutation in a single gene is the cause for the disease: the gene RB1 on chromosome 13 (Mastrangelo et al. 2008). Since this form of cancer was fatal up until recently, these mutations were ruthlessly weeded out. Death meant that the whole genome of individuals with retinoblastoma was destroyed. The cooperation between genes broke down because of an error in a single gene that consequently didn't function collectively with the other genes.

In other parts of the human genome almost anything can happen without having an effect. Here, mutations are common and the control mechanisms are more relaxed. For example, there are long stretches of DNA that seem not to provide instructions for anything. In one experiment, a whole percent of the genome in mice was removed, without any noticeable difference (Nobrega et al. 2004).

Mutations and other copying errors happen all the time. Sometimes the error results in cooperation breakdown. Then that cell line ends, self-copying stops and death ensues. When the system is fully developed it is almost impossible for a gene to make it on its own—to exit its context and carry on as a stretch of loose DNA.

Not that breaks in cooperation never happen; they happen worryingly often. Researchers have found genes that copy themselves into several places in the chromosomes, genes that jump from one place to the other, and genes that become represented with more copies than other genes at self-replication (Hooekstra 2003). But there also exist genes that code for strict control mechanisms against that form of 'cheating'.

Genes must also function together to prevent providing conflicting instructions for the construction of the organism. Genes that provide superior hunting capabilities (flexible limbs, soft pads, and an ability to predict the movement of the prey) must cooperate with genes that are good for the metabolism of meat. Genes that result in a craving for grass function better together with genes that result in molars and the ability to chew cud.

Cooperation between genes can be said to be the 'innermost' form of cooperation in a human body. The collective functioning of genes is what makes life possible and may even be the best definition of life itself. There are, however, different solutions to how one group of cooperating genes are separated from other groups of cooperating genes.

Simple Cells—Prokaryotes

The simplest life with DNA or RNA that completely runs its own reproduction are single-celled organisms without a nucleus: the prokaryotes, or—as they are termed now—bacteria and archaea.[1] It is from this stage we have the first evidence of life, from at least 3.5 billion years ago (Schopf et al. 2002), though new evidence pushes this date as far back as 4.1 million years ago (Bell et al. 2015).

Within cells of this kind, all parts work together to run the cell's machinery in the same way as the cell's genes cooperate in order to be copied. In this type of life, each cell is its own individual.

The genes of prokaryotes are collected in a single ring molecule—a nucleoid. These genes not only contain copying instructions but also instructions for the cell's metabolism and maintenance. Genes that carry good instructions for a well-functioning cell machinery get copied to a higher extent than genes that carry instructions for a less well-functioning machinery. But it is equally true that a cell with better functioning cell machinery replicates more often than a cell with less well-functioning cell machinery. The cell is in this way also a replicator, a continuation of the genes. The 'interests' of the genes and the 'interests' of the cell concur—though not completely.

For example, bacteria sometimes exchange genetic material. The process starts with the outgrowth of a small spear-shaped extension from one bacterium into the next. Then genes traverse the outgrowth and are in this way passed on from one cell to the other. Not all genes, though; genes that are central for keeping the bacterium alive stay put. One example of genes that can move like this is the type that provides bacteria with resistance to antibiotics.

What the recipient bacterium has to gain from the transfer is obvious: it gains some genetically coded trait and can in this way survive and keep making copies of itself. But why does the other bacterium send over the genes—what is in it for the sender?

This mechanism is more easily understood if you change perspective and investigate what advantage there is for the genes instead of what advantage there is for the bacteria. An 'ordinary' gene—a gene that does not jump between individuals—spreads only through one process: through cell fissioning from one cell generation to the next.

But a gene that also carries instructions to be contagious from one cell to the next has more reproductive potential than this, because such a gene can reproduce and be copied in two ways: through cell fission and through contagion. It is not so hard to see that a gene that gets copied in two ways becomes more numerous than a gene that gets copied in only one way.

[1]A key difference between bacteria and archaea is that the cell wall of archaeans consists of pseudomurein while the cell wall of bacteria consists of peptidoglykan.

Thus, what looks as self-sacrificial cooperative behavior turns out to be nothing else than the product of selfish genes—a form of beneficial genetic 'disease' that spreads through populations of bacteria. In this way, selfishness on a genetic level becomes something that looks as costly cooperation on an individual (cellular) level. The unit that has the benefit from the genetic donation is not the donating bacterium itself, but the genes that carry the instructions for antibiotic resistance.

This is a common theme in many types of cooperation, that to understand what is going on, it is crucial to change perspective from individuals to replicators. Cooperation on one level (individuals) has its foundation in an advantage for replicators on another level (genes).

To understand cooperation one must therefore constantly ask the question that is the title of this book, 'For whose benefit?'—*Cui bono*? This will become even clearer on the next level, where several bacteria gang together and life enters a more entangled stage—cells with nuclei.

More Complex Cells—Eukaryotes

In 1967 a sensational scientific article was published with the title 'On the origin of mitosing cells' in the *Journal of Theoretical Biology*. The author was a young faculty member in the Department of Biology at the University of Boston, Lynn Margulis (Lynn Sagan at the time, as she was married to the famous astronomer Carl Sagan) (Sagan 1967). The idea she presented was so controversial that according to her own story the manuscript was 'rejected by about fifteen scientific journals' before being accepted for publication (Brockman 1995).

Margulis had authored a milestone article where she presented a theory that at first sight seemed almost wacko, but eventually proved to be correct. The article outlined the evolution of cell organelles[2] within the type of cells that carry the name *eukaryotes*. Where the term prokaryote comes from the Greek *pro* = before and *karyon* = nucleus (i.e. before the nucleus) eukaryote instead has the prefix *eu* = true. It is thus the presence of a true cell nucleus that gives eukaryotes their name.

While prokaryotes (bacteria and archaea) have their genetic material collected in a ring, eukaryotes have their genetic material collected in chromosomes. These lie protected within an extra cell membrane inside the cell: the cell nucleus. Multicellular animals are eukarytes, as are multicellular plants and fungi—we all have an extra cell membrane that protects our DNA within (almost) all cells in our body.

But not only that, we also have *mitochondria*. These are the energy factories of our cells. Within eukaryote cells, mitochondria convert glucose and oxygen to adenosine triphosphate, or ATP, the molecule used as energy source in most energy-demanding processes in cells. In addition, plants also have *chloroplasts* containing chlorophyll. This is where photosynthesis occurs.

---

[2]The cell's 'organs'.

So where do cell nuclei, mitochondria and chloroplasts come from? How did they evolve? Interestingly, all three contain DNA. While the cell nucleus seems to be the evolutionary result of the cell's own cell wall folding in on itself and detaching, Margulis suggested a radical proposal for other organelles. Many bacteria gain nutrients through absorbing material around them. Research has now confirmed Margulis' hypothesis, that the organelles were once bacteria ingested by another bacterium without being digested. Instead, the eaten bacteria were able to carry on living and reproducing within the other bacteria; they were 'symbiotically acquired' (Sagan 1967).

These intruder bacteria could make their living on stuff that the host bacteria ingested, or on free-flowing detritus within the host. The tight interdependence that slowly emerged is what in biology is termed *symbiosis*—a close, mutually dependent, long-term relationship between different species. Over time, roles crystallized—the symbiotic bacteria specialized. Some of the intruder bacteria became energy factories: the mitochondria. In plants, chloroplasts evolved in a similar manner.

Mitochondria and chloroplasts are of special importance to evolutionary biologists as they carry their own DNA. At cell fission, some mitochondria and chloroplasts follow one half of the cell while some follow the other. This means that two genetic inheritances are transferred in each cell fission (or three, in plants, through the chloroplasts).

The unique thing about mitochondria and chloroplasts is that they are transmitted only via egg or ovules of a flower, never via sperm or pollen. This means that their DNA is transmitted only through the female lineage; this inheritance is totally maternal.

Interestingly, some plants have mutations in their mitochondria that have resulted in the termination of cooperation with the rest of the organism. Since mitochondria are transmitted only on the female side they derive no necessary benefit from the squandering of resources on males—so these mutations in the mitochondrial genome result in sterile males. This is another example of a mutation resulting in a gene in conflict with the rest of the genes in the organism, resulting in cooperation breakdown.

The mitochondria present in human cells means that we are related to bacteria along one line and through another to archaea. Foremost, the relationship is through the original evolutionary line of the DNA in our cell nuclei, but we also carry a separate relatedness through our mitochondria—but only on our mother's side.

Recent developments indicate that the story of bacterial endosymbiosis is even stranger than we thought—some eukaryotes seem to have acquired their chloroplasts by ingesting not another photosynthetic bacterium, but another eukaryote who already had gained a chloroplast through ingesting a photosynthetic bacteria. Even more complicated, some dinoflagellates (a type of plankton) have acquired their chloroplast by eating some of *those* eukaryotes. Further, some reef-forming corals contain this type of dinoflagellates *and* some of the photosynthetic bacteria that were ingested by the 'first' ingesting host bacteria (Godfrey-Smith 2009).

It is not currently known exactly when the first eukaryotes emerged, the current estimate is sometime around 1.5–2.5 billion years ago. Free-living eukaryotes still exist; we know them as protozoa, one-celled animals, and cyanobacteria, one-celled plants. But eukaryotes have also evolved into such remarkable creatures as fungi, plants and—not the least—animals as ourselves.

Thus, (almost) each and every cell in our bodies consists of cooperating cells within cells, specialized units that from the beginning were separate organisms (and thus an example of egalitarian cooperation—cooperation between unlike parts) but now have lost their autonomy and are so integrated that they are indivisible units. Their internal cooperation is for the benefit of all the cells and their genes.

As before, the benefit of this cooperation is assessed through the number of genes present in the next generation. Genes that result in traits that give cells and their mitochondria an edge will over time be represented more numerously. One such successful way to become more numerous has been for the cells within cells (the eukaryotes) to cooperate with each other.

Note that the emergence of eukaryotes is a coming-together of different bacterial species to a new type of organism, where the component parts eventually become so interdependent that they cannot live without each other. This is not the last time we will see examples of this kind, think for example of the fungi and algae that make up lichen. What exactly constitutes an 'organism' can therefore be difficult to sort out (Queller and Strassmann 2009). Because evolution occurs gradually, there exist blurry phases along the way from separate organisms to one organism.

Multi-cellularity

The next layer of cooperation within humans is multi-cellularity: cooperation between cells. Cooperation between cells in general exists in two flavors: cooperation between cells with *different* genetic material (egalitarian cooperation) and cooperation between cells with *the same* genetic material (fraternal cooperation) (Queller 1997). Humans are (mainly) multi-cellular in the latter sense, so we will focus on that now: a form of cooperation that has evolved more than 25 times whereof three have led to the evolution of more complex organisms: plants, fungi and animals. Cooperation among entities with *different* genetic material will be discussed later.

Bacteria that replicate through fission will ultimately result in a large number of cells with the same genetic material gathered in a heap of many cells. But these cells all look the same, function the same and don't depend on each other for their existence; they are a colony of cells, not one multi-cellular organism. True multi-cellularity is instead when many cells from a single origin form a distinct unit where cells depend on each other for the survival and reproduction of the larger unit.

Cells in fungi, animals and plants are specialized—different cells carrying the same genetic information having different functions. During individual development, each cell receives signals from its surroundings resulting in genes turning on or off which in turn is what causes specialization. This process ensures that the right organ ends up in the right place—a process completely guided by the interaction of the genetic material with its immediate surroundings.

This is how we get heart, lungs, muscles and a brain; this is how trees end up with trunks, branches, twigs, roots and leaves. The specialized parts carry out specialized tasks for the benefit of the whole organism. Despite their marked differences, all these parts are built of cells possessing the same genome. The differentiation into organs is entirely caused by differences in the signals the cells receive depending on where they happen to be in the body. Understanding this process is a hot current research topic.

Just as was the case with the genes, the cells' specializations open up for new possibilities of evolving complexity. Without specialization you just end up with many similar cells stuck together doing the same thing. With specialization, new forms of organisms can be built; big bodies consisting of specialized organs.

One spectacular consequence of specialization in multi-cellular organisms is that because all cells have the same genetic material, genes can now code for pre-programmed death—cell death. As long as all genes are copied effectively from generation to generation, the survival of single cells is of no consequence. One consequence of multi-cellularity is therefore that death becomes built into life itself.

Many structures in multi-cellular organisms actually consist of, or are the result of, dead cells. The discovery of programmed cell death resulted in the 2002 Nobel Prize in medicine for the South African Sydney Brenner, American Robert Horvitz and Briton John Sulston.[3] One of the most extreme examples of cell death is the wood of trees. Wood consists of cells that have grown and then been instructed by their genes to die. The overwhelmingly largest proportion of a tree thus consists of dead cells.

Even humans are constructed in this way. For example, the outermost layer of human skin consists of 25–30 layers of dead cells that have also been programmed to die. The human skeleton, however, instead consists mostly of living cells, even though the cavities in the skeleton are the result of programmed cell death.

Cell death is an important part of human development all the way from the fertilized egg to adult being. For example, without programmed cell death you wouldn't have separate fingers and toes. During fetal development, fingers are stuck together on each hand and toes are stuck together on each foot. Only later do cells between the fingers and toes die so that these become are separated. If something goes wrong in this process—which isn't entirely uncommon—you are born with fingers or toes stuck together.

[3]The Nobel Prize in Physiology or Medicine 2002 was awarded jointly to Sydney Brenner, H. Robert Horvitz and John E. Sulston 'for their discoveries concerning 'genetic regulation of organ development and programmed cell death'.

Cooperation between cells that have the same genes means that 'individual' is no longer synonymous with 'cell'. Instead, individuals now consist of a collection of cooperating cells. Australian philosopher Peter Godfrey-Smith has termed this process 'de-Darwinizing', where individual cells of a body no longer compete with each other, having lost their autonomy they instead cooperate together for the good of the individual (Godfrey-Smith 2009)—or for the good of the common genome, depending on perspective. The same process has occurred for organelles (mitochondria and chloroplasts) and genes.

It is one of these collectives of specialized cells you see when you look at yourself in the mirror.

Cooperation between cells with the same genes is also what has led to aging and death. In single-celled organisms death also exists, of course, but not as a programmed component of life. All death in single-celled organisms is due to accident, not design. Bacteria and other single-celled organisms are potentially immortal; they just go on fissioning and have done so since the dawn of life. From the first bacterium to all bacteria that exist today there is an unbroken chain of cell fission where no 'new' cell is the descendant of another, but largely the same cell in two versions. Mutations occur in individual cells and are transmitted to later copies through duplication.

Multi-cellular organisms function according to a different principle. Here, most cells have as their task to maintain life in the whole collection of cells; the whole organism. Only a select few cells have reproduction as their main task (in humans, these are sperm and eggs). If mutations happen in the cells that are *not* involved in reproduction, these are consequently *not* transmitted to the offspring.

Almost all multi-cellular organisms go through a single-celled stage when reproducing, often by combining their genetic material with that of another individual in the process we term sexual reproduction. We are all single-celled, initially. If you want to reverse the perspective, you can thus say that our bodies are really a very roundabout way of single-celled organisms to get together to reproduce. Or, as the British satirist Samuel Butler phrased it already 1877: 'A hen is only an egg's way of making another egg.'

As long as reproduction has occurred, all other cells of the body are unimportant and can be discarded, which is exactly what happens. After our reproductive age, our bodies start slowly malfunctioning, a process slowed only by the advent of modern medicine. In the end, all multi-cellular organisms die, leaving only their offspring behind to go through the same process. Evidence points to all living things on earth coming from the same initial cell; all living forms represent genetic lineages that have not yet died out. Though our bodies are mortal, our genetic lineages are potentially not.

The single-cell phase, that takes place just after fertilization, ensures that all cells in the resulting body will have the same genetic material. Thus, individual cells from the body will have little incentive to break cooperation and leave to live on their own to the detriment of the whole. If such a mutation occurs in a subsequently developed body cell, it is not transmitted to the next generation.

But this is a truth with caveats. For example, many plants can reproduce through saplings. It can be argued that this is a form of growth, rather than reproduction, even though unity is broken with the parent organism when the sapling loses contact with the parent tree. Such reproduction through saplings is similar to bacteria fissioning—a splitting of organisms where the offspring is genetically identical to the parent (almost—there are always a few mutations).

Peter Godfrey-Smith has pointed out that there exist three varieties of reproduction, basic, collective and scaffolded. Fissioning of bacteria and through saplings is what can be termed 'basic', whereas our type is collective in that it requires more than one individual for reproduction to succeed through sexual reproduction. Finally, viruses, chromosomes, mitochondria and 'memes' (we will get to them later in the book) reproduce by high-jacking another system and thus require the existence of another structure than themselves to reproduce.

Nevertheless, it is still overwhelmingly common that a unicellular phase exists somewhere in multi-cellular organisms' life cycles (also for trees that can reproduce through saplings). For people this is fortunate, otherwise mutations would have accumulated in the genome at terrifying speed.

We humans have a mutation frequency of about 0.000000025 mutations per nucleotide (one nucleotide is a 'letter' in a string of DNA) (Nachman and Crowell 2000).[4] This means that each person leaves about 175 new mutations to each child. However, most mutations are completely harmless and pointless.

In 1961 the American anatomist Leonard Heyflick presented results indicating that human cells can handle about 40–60 cell divisions before they stop fissioning (Hayflick and Moorhead 1961). At the end of the life cycle of a cell that has gone through about 60 divisions, around 10,000 mutations ($60 \times 175 = 10{,}500$) have thus accumulated. What stops further cell divisions and causes the death of the cell is that the ends of the chromosomes—the so-called telomeres—accumulate mutations and become shorter and shorter.

Would continued cell division be desirable? If we could in some way reconstruct the telomeres, wouldn't we be able to stop suffering and death?

There are actually some cells that produce a substance called telomerase that achieve exactly this. Telomerase allows some cells to divide indefinitely. We call these cells *cancer*.

The infinite replication of cancer cells has the consequence that more and more mutations accumulate as they divide. Cancer cells thus function less and less as the initial cells from which they developed and instead become more and more defective. Researchers are consequently now investigating the possibility of medicating cancer by terminating the production or function of telomerase.

Thus, cancer is nothing less than cells that have broken cooperation because the genes that exist to control the cooperation of single cells or control the copying of

[4]More precisely, mutation rate is between 1.3×10^{-8} and 3.4×10^{-8} mutations per nucleotide (a nucleotide is a 'letter' in a DNA string). This means that every person bequeaths approx. between 91 and 238 new mutations to every child.

genes have been destroyed or have mutated. Even if single genes are merely components of cells, and single cells are merely components of multi-cellular individuals, cancer functions as a drastic reminder that we really are collections of cells.

Interestingly, cancer in some organisms is not equally detrimental as in animals. In trees, for example, cancer can emerge and kill a part of the individual but in general seem to be not nearly as harmful as in animals like us. Part of the reason is that cancer cells cannot spread around the tree due to their more rigid cell walls. Thus, problems at one site doesn't necessarily end in the death of the whole organism.

It is a part of the life cycle of healthy cells in humans to die after about 40–60 cell divisions, for the benefit of the whole organism (and the organism's genes). In this way, the body is purged of accumulated mutations. However, certain cells exist that do not accumulate mutations: stem cells—cells that yet haven't been specialized to become, for example, liver cells or eye cells. These exist in different places in the body and continuously replace dead cells and in this way function as the body repair system. How stem cells can keep dividing without accumulating mutations is another really hot research topic (Nature Reports: Stem Cells).

Since genetic information from the body at large is not transmitted to offspring, sexual reproduction is one protection against 'selfish' cells having lost their ability to cooperate. But sexual reproduction also has another advantage (ok, two), in that beneficial mutations from several independent individuals through this form of reproduction can be combined into a single genome.

How does this work? Imagine a situation where a good adaptation occurs in one individual and another good adaptation in another individual. In organisms without sexual reproduction these two adaptations will never meet. In organisms with sexual reproduction, on the other hand, individuals can mate and have offspring carrying both adaptations. Sex is thus a form of cooperation between two individuals with different genomes that speeds up the evolutionary process. We will get back to why and when sex evolved in a later chapter.

As we have seen before, bacteria have instead solved this problem through gene donations. Antibiotic resistance can thus be seen as a form of bacterial sex.

The death of a single cell is no catastrophe for a multi-cellular organism, whether death is pre-programmed or random (if you happen to cut your finger, for example). Your existence does not depend on the survival of some 'key cell' or 'executive cell'. On the contrary, your body can lose many cells due to natural processes or random cell death without this killing you.

There is simply a surplus of cells, which makes your body very robust in relation to the individual cells. This surplus makes the whole more resilient than if you had been built by just enough cells to run a functioning body. This is a common theme where cooperation between units has developed into a new, larger entity—functioning cooperation in nature isn't trimmed and lean, but instead characterized by abundance, resulting in truly impressive resilience.

Further, organisms consisting of several cells represent a higher degree of order. When a multi-cellular organism grows from being one cell to being adult, the building blocks (molecules) are sorted successively according to a bauplan coded in the genes of every cell, the cells themselves ordered into a multi-cellular unit—an individual. More and more matter is bound in ordered structures on each level of cooperation in our body—in the genes, in the cells and in the body.

The Austrian physicist Erwin Schrödinger formulated it thus: 'The essential thing in metabolism is that the organism succeeds in freeing itself from all entropy[5] it cannot help producing while alive.' Organisms liberate themselves from entropy by absorbing order from the environment. Animals and fungi do this by ingesting once or currently living things, which they digest and excrete in a more disorderly state. Plants do it through transforming order in sunlight into chemical bonds (and heat).

Multi-cellular organisms where all cells have the same genetic make-up is the type of organism that we commonly notice around us. Humans, ants, plants, fungi—all large-scale life is multi-cellular. Individuals in these groups are composites of a large number of cooperating parts—cells—where life and death are unimportant for the single cells as long as the composite individual lives on.

Such steps of ever higher-level cooperation have been termed 'major transitions in evolution' in an important book by the evolutionary biologists John Maynard Smith from Great Britain and the Hungarian Eörs Szathmáry (Maynard Smith and Szathmáry 1995).[6] They postulate that many traits are in common in each transition. For example, each level of increasing cooperation is characterized by the specialization of common units.

It is thus not *more of the same* that is the explanation of how the whole can become more than the sum of its parts, but *more specialized units*. When the building blocks have been specialized, they most often lose the ability to multiply on their own, outside the cooperative context.

On each higher level of cooperation there are control mechanisms to keep the parts from breaking unity through behaving selfishly to the detriment of the whole. On each level there is also redundancy of constituent parts that makes the whole resilient to the loss of single parts. The picture that emerges is that our bodies are like series of Russian dolls where each higher level of cooperation includes another, more fundamental level of cooperation, all the way down to the self-replicating molecules: the genes—the replicators.

But incredibly, our bodies consist not only of cells with our own genetic code. There is much more to us than meets the eye in the mirror.

---

[5]Disorder.

[6]They include a few more steps, principally at the outset of life, which are unimportant for the argument since those steps are not part of the cooperation that makes up the human body or the 'id', and do not include the 'ecosystem step' that I describe above.

Mobile Eco-systems

It is estimated that about 90% of the cells in our body are not our own, but belong to bacteria, mostly in the gut. About 100,000 billion bacteria of 36,000 species (Frank and Pace 2008) live there, together weighing about 1.5 k. Some of these exist in the stomach and the small intestine, but most live in the colon. The number of different *types of genes* that exist in these bacteria is more than ten times as numerous as our own: approximately 20,000 genes (Arumugam et al. 2011). Counting the *total number of copies of genes*, we carry more than 100 times the number of bacterial genes as our own.

To an extent, this incredible diversity defines us; it turns out that humanity is divisible into three different groups—'enterotypes'—depending on the bacterial ecosystem in our gut microbiome—there are only certain mixes of bacteria that result in stable internal eco-systems (Child and Macfarlane 2008). These bacteria are welcome and necessary guests in our bodies as they cooperate with us in exchange for room and board.

Normally, these cooperating bacteria out-compete disease-causing bacteria. You notice this if you ever ingest broad-spectrum antibiotics—the type that indiscriminately kills many types of bacteria—because by doing so you partly weaken your natural protection. It is therefore not surprising that one of the common side effects of antibiotic-treatment is stomach problems.

It is estimated that 5–30% of the nutrients our bodies absorb are not derived from the food itself, but are instead waste products from our friendly bacteria-symbionts who are the ones actually processing some of the food we eat. We also get K-vitamins and other necessary 'waste' products from bacteria in this way. This is why newborns in many countries are given K-vitamin supplement. The gut flora of babies takes a while to become established.

Bacteria that produce waste products that the body doesn't want are of course not welcome. Nor are viruses that move into our cells and hijack them from within to make more viruses (this is how viruses 'live'). Our immune systems are constantly busy with these intruders and trying to get rid of them.

'Our' bacteria co-habit with us because it is beneficial for them—they get both room and board. Our immune system doesn't attack the 'good' bacteria since our bodies are better served by having them in place than not having them there; a mutual dependency that both parties derive benefits from. A kind of *symbiosis*, actually, since we can't live without them and they can't live without us. Cooperation where all parties gain benefits is also termed *mutualism*.

But can we really count these bacteria as part of our bodies, our physical selves? If we include them this would be our third form of kinship with bacteria; our first alliance is through the genes in our cell nuclei, the second via the genes in the mitochondria (the energy factories of the cells) and now the third form of kinship via the large number of bacteria that we are totally dependent on inhabiting our guts. Including them would immediately increase the number of human genes by two orders of magnitude. But no, we currently count as interdependent species of organisms, not the same species.

There are many other stowaways on and in the human body, some that we would rather be without: tapeworms in the intestines, lice, fleas, ticks and other small pests. These are not examples of cooperation but are instead an example of another type of interaction where only one of the parties derives a benefit: *parasitism*. Note, however, that a parasitic relationship can be symbiotic—symbiosis does not only refer to cases where with parties benefit.

But what about the organisms that live on and in us, but neither benefit nor harm? There are, for example, two species of small arthropods (*Demodex folliculorum* and *Demodex brevis*) that live only in human eyelashes, without causing any ill symptoms whatsoever. About half of all humans carry them; they are a bit rarer in children and a bit more common in the elderly. They use us only as substrate and food source without providing any benefit or costing us anything. This is an interaction termed *commensalism*.

From this it should be evident that a human being is not just a collection of cooperating cells with common DNA, but rather a mobile eco-system. Every human is a sort of super-organism with one overarching common evolutionary goal: everyone striving to reproduce; all genes to get copied.

Do you see this collective, this layered eco system, when you look at yourself in a mirror? Or does your self stubbornly persist in presenting a single unified individual? If you still cannot help but see yourself as a single unified physical individual then don't worry—it gets worse. This far we have only considered your *physical* self. But what does your *psychological* self consist of—the self that discerns 'its own' mirror image?

References

Arumugam, M., et al. (2011). Enterotypes of the human gut microbiome. *Nature, 473*, 174–180.

Bell, E.A., Boehnke, P, Harrison, T.M., & Mao, W.L. (2015). Potentially biogenic carbon preserved in a 4.1 billion-year-old zircon. PNAS 112: 14518–14521.

Brockman, J. (1995). *The third culture: Beyond the scientific revolution*. New York: Simon & Schuster.

Child, M., & Macfarlane, G. (2008). The human ecosystem. http://www.medscape.com/viewarticle/584833

Dawkins, R. (1978). Replicator selection and the extended phenotype. *Zeitschrift für Tierpsychologie, 47*, 61–76.

Frank, D., & Pace, N. (2008). Gastrointestinal microbiology enters the metagenomics era. *Current Opinion in Gastroenterology, 24*, 4–10.

Godfrey-Smith, P. (2009). *Darwinian populations and natural selection*. Oxford: Oxford University Press.

Hayflick, L., & Moorhead, P. S. (1961). The serial cultivation of human diploid cell strains. *Experimental Cell Research, 25*, 585–621.

Hooekstra, R. F. (2003). Power in the genome: Who suppresses the outlaw? In P. Hammerstein (Ed.), *Genetic and cultural evolution of cooperation* (pp. 257–270). London: Dahlem Workshop Reports, MIT Press.

Mastrangelo, D., De Francesco, S., Di Leonardo, A., Lentini, L., & Hadjistilianou, T. (2008). The retinoblastoma paradigm revisited. *Med. Sci. Monit.* 14: RA231–240.
Maynard Smith, J., & Szathmáry, E. (1995). *The major transitions in evolution*. Oxford: Oxford University Press.
Nachman, M. W., & Crowell, S. L. (2000). Estimate of the mutation rate per nucleotide in humans. *Genetics, 156*, 297–304.
Nature Reports. Stem cells. Retreived December 2016 from http://www.nature.com/stemcells
Nobrega, M. A., Zhu, Y., Plajzer-Frick, I., Afzal, V., & Rubin, E. M. (2004). Megabase deletions of gene deserts result in viable mice. *Nature, 431*, 988–993.
Queller, D. C. (1997). Cooperators since life began. *The Quarterly Review of Biology, 72*, 184–188.
Queller, D. S., & Strassmann, J. E. (2009). Beyond society: The evolution of organismality. *Philosophical Transactions of the Royal Society B: Biological Sciences, 364*, 3143–3155.
Quote Investigator. Retreived December 2016 from http://quoteinvestigator.com/2011/02/09/darwinism-hope-pray/
Sagan, L. (1967). On the origin of mitosing cells. *Journal of Theoretical Biology, 14*, 255–274.
Schopf, J. W., Kudryavtsev, A. B., Agresti, D. G., Wdowiak, T. J., & Czaja, A. D. (2002). Laser-raman imagery of earth's earliest fossils. *Nature, 416*, 73–76.
Vaidya, N., Manapat, M. L., Chen, I. A., Xulvi-Brunet, R., Hayden, E. J., & Lehman, N. (2012). Spontaneous network formation among cooperative RNA replicators. *Nature, 491*, 72–77.
Williams, C. G. (1992). *Natural selection: Domains, levels, and challenges*. New York: Oxford University Press.

Your Psychological Self

Abstract Cooperation between neurons is described, giving rise to our "psychological selves". The process according to which the brain grows is outlined. It is made clear that parts can be removed from the brain without removing our core "selves". Is there even a core? Also, the point is made that the brain only is our "hardware" and that the discussion of our "software" in later chapters will be crucial for a more complete picture.

> Of course it is happening inside your head, Harry, but why on earth should that mean that it is not real?
>
> Albus Dumbledore, from Rowling:'s Harry Potter and the Deathly Hallows (2007).

If your physical body is a wonder of cooperation, your psychological self more resembles a miracle.

A human brain consists of about 86 billion neurons (Azevedo et al. 2009). It is sometimes popularly stated that this is about equivalent to the number of stars in our Milky Way galaxy, but this isn't true: the Milky Way has about twice that many stars. However, unlike the stars, each of the brain's neurons is in direct contact with about 10,000 other neurons. If you multiply these two numbers you find that an adult human brain contains approximately 860,000,000,000,000 (860 trillion) connections.

This network structure holds the secret to brain function. Researchers have learnt to construct small artificial networks built on similar principles for solving problems difficult to solve using conventional methods. Just like the human brain, these artificial networks have to be trained to solve tasks, but once trained, they can be used for tasks such as pattern recognition, signal processing and scheduling.

The brain's enormous network of neural connections emerges from a complex combination of determinism, randomness and simple rules. During growth, nerve cells send tentacles in every direction. This growth process is genetically governed, but only to a degree. Genes, for example, are part of the explanation for why specific parts of the brain are organized in specific ways for solving specified tasks.

P. Lindenfors, *For Whose Benefit?*, DOI 10.1007/978-3-319-50874-0_3

Neuronal growth is not genetically governed in detail, however, but through generalized 'grow-and-send-out-tentacles' instructions. This growth is guided by chemical signals that pull the neurons toward receptors in some nerve cells while repelling them from others. The network of neurons thus seems organized through very simple rules intrinsic to each neuron, similar to how ant paths emerge around an ant nest through very simple behavioral rules intrinsic to each ant.

Just as with ant paths, connections that are then used more often are strengthened while connections that are not used are pruned. This process of 'neural Darwinism' (Edelman 1987) determines which connections become permanent and which are discarded.

(Bonus fact: Alcohol disturbs both the genes providing the general grow-and-send-out-tentacles instructions as well as the signal substances that tell nerve cells what direction they should be growing in. A brain in development is extremely sensitive to alcohol.)

The neural Darwinism that culls the less useful from the useful connections is a process that continues for a long time during development. It is only in the late teens that the process nears completion in the frontal lobes, where judgment and executive decisions are handled (no surprise to those of us who have survived our teens).

The process continues through life. This is how the brain responds to training; this is how memories are stored. So don't forget to encourage the connections you want to keep.

But how does this miracle brain work? A large part of what makes humans unique is localized to the brain, but despite this we lack a fundamental understanding about some crucial parts. Where, for example, does the subjective experience of an 'I' come from; the 'I' that is experiencing the world—the most central part of our experience?

A common way to view this 'I' throughout history has been to assume that there is some form of non-material substance that lives within the body or the head: a soul. But this is a problematic explanation for several reasons.

If 'soul' implies the existence of something non-material, yet existing and changeable. The introduction of such an entity would make the possibility of investigating consciousness through scientific enquiry impossible. The problem with assuming the existence of a non-material soul is, simply put, that it is non-material—literally like looking for a ghost.

Allowing non-material explanations—untestable explanations—means that *anything* becomes permissible as an explanation. Do we really have a soul? Or are we Thetans governed by the evil Lord Xenu? Are we shards of a great, collective consciousness? Is our consciousness an inter-dimensional slug, remotely controlled from one of Jupiter's moons? All these explanations are equally impossible to investigate scientifically—and thus just as worthless.

If this sounds strange and a bit unsettling it is because you grew up in a society where the idea of something immaterial that survives the body has had a central place in intellectual discourse for millennia. But invoking an immaterial soul as an explanation for the 'I' carries with it problems if you want to stick to the scientific world view. (And if you reject the scientific world view you are reading the wrong book.)

Another problem with the idea of a non-material soul is that the brain and the rest of our bodies are material. How would a non-material soul communicate with a material brain? If the non-material soul can 'touch' the material brain with some part, then that part of the soul has to be material—otherwise there is no 'touching'. But if this is the explanation, then how do the immaterial and the material parts of the soul communicate? The problem has just been shifted, not solved. If the brain has some part that can receive immaterial 'touching' from an immaterial soul, then that part of the brain must have an immaterial part that can touch the material part. How would that communication happen?

However you view the problem, at some point, somewhere, something immaterial has to 'touch' (communicate with) something material. Thus, if you include a soul in your reasoning, this assumes a miracle at each thought, a somewhat unsatisfying proposition for a scientific explanation.

Do not, however, confuse this dismissal of an immaterial soul with the complete dismissal of the existence of immaterial entities. For example, mathematics is often said to exist in a realm unconstrained by material reality. Concepts of imaginary numbers and multi-dimensional algebra are possible to hold and use, even if they do not 'exist' in the material sense of the term—and surely such ideas can change our brains—this is how we learn.

This is, however, a completely different form of immateriality than in the case of the soul, which is supposed to transmit thoughts, feelings and awareness even without the body. In mathematics, each stage of learning, understanding and using the information involves encoding information in a material structure—the brain—much as mathematics is encoded in a computer. Thus the material (our brain) gives rise to and uses an immaterial reality (mathematics). Perhaps the immaterial 'self' emerges from the material in a similar way?

A Soulless Existence

So we cannot postulate the existence of a supernatural soul, at least not scientifically. We can, however, speak of a 'self'.

All (most?) of us experience that there is a 'self' inside the head somewhere. You can evoke this experience through a thought experiment. Look around—take a moment, experience your surroundings. Then close your eyes and move around in the memory of what you just experienced.

I am currently sitting in a room with windows to my right, two messy bookshelves on either side of a fireplace straight ahead, and a sofa to my left. When I close my eyes I can focus on the window and to a certain degree remember what was on the windowsill and if it was light or dark outside. If instead I focus on the sofa, I remember its color and if someone was sitting there or not. I can remember if there was a fire in the fireplace or not. All this is reachable through inspecting my memory, through my 'self' moving around in my memories.

But what am I doing, really? And above all, who is doing the doing? The American anthropologist and language researcher Terrence Deacon has pointed out the irony in this introspective way of concluding that there exists a 'self' (Deacon 1997). What you really are doing is shutting out reality and contemplating a virtual reality—a mental construct that the brain makes for you. It is not reality but a simulated version that confirms your experience of a self.

But if I can explore a mental image of the world, where is this mental image situated? Surely, it must exist in my head and as such be a part of my self. But if you agree on that, that the memory is part of the self, what is it then that does the exploring? Another part of the self? What *is* a part of a self?

If you do not accept that the memory is a part of the self, what is it that differentiates the memory from the self? Is there some border inside the head that divide the outer service parts from the inner self-parts?

The American philosopher Daniel Dennett (1992) has pointed out that if the latter is true, the following should be possible: follow a neuron from the eye into the brain, to the visual center. A first culling of the visual input takes place in this visual center, automatically, outside of conscious control. Movement is prioritized and generates an extra strong signal inward. A collection of visual inputs that the brain has learnt to associate with food and danger also provide extra strong responses. These impulses make their way up towards the forebrain which identifies what you are seeing as, for example, a piece of chocolate. Would you like a piece?

In that case you may attempt to grab it. Similarly, we should now be able to follow the neuronal impulses that start the movement of your hand backwards. Follow the neurons destined for the muscles controlling the arm and you land in the brain's motor centers. Impulses instructing the brain's motor center to move the arm download from other parts of the brain.

Somewhere, nerve impulses that go into the brain have to meet the nerve impulses that go out. Exactly on that spot logic would place the self, that very part of the brain that makes up its mind and decides to pick up the piece of chocolate and move it to the mouth. Exactly at the division between impulses going in and going out must be the seat of that elusive 'you', a unit that watches visual input and then reacts by activating impulses to the muscles.

But how would such a self function? It also receives impulses going in (looking at images of the chocolate) and then sends impulses going out (by initiating movement to pick it up).

If this system sounds familiar, it is because it is exactly the type of system to manage information and reaction that we tried to explain in the first place! We followed ingoing visual impulses and outgoing motor impulses, trying to find a center, to place a self that now, apparently, also has to receive and send out impulses.

Will we find yet another brain inside this new system that receives the incoming nerve impulses and gets outgoing impulses sent in response? Is there yet another self inside that unit demanding an explanation itself? How does this even smaller self work? We keep moving to ever smaller units.

This is a so-called homunculus-argument, explaining the functioning of the brain by assuming an internal observer with a brain of its own. And if we want to explain the workings of the observer's brain in the same way—it worked so well the first time—we end up with an even smaller observer with an even smaller brain. In the end, we end up with an infinite number of brains within brains.

It has to stop somewhere. Somewhere we should end up with the innermost neuron—an 'executive neuron'—which functions as a main switch that sometimes 'decides' to react on an incoming nerve impulse by initiating an out-going nerve impulse, and sometimes doesn't. But how would this 'executive neuron' make up its mind?

Majority Rule

This reasoning obviously doesn't work. And if you actually look at how the brain functions it doesn't have any empirical backing at all. Instead, large networks of interconnected neurons turn on and off and through this collective action activate other networks of neurons. There are millions of 'executive cells', that together 'make up their mind'.

These neurons, acting together, turning on and off, are your 'innermost' self. The brain makes up its mind through a form of majority rule, making the brain an enormous cooperating unit. No executive is watching inside your head. The brain (and your body) *is* yourself. There is no executive neuron; instead there are many nerve cells together that form 'the executive'—that form your 'self'—and where to draw the border between the 'outer' and the 'inner' self is not at all clear.

The British Nobel Prize winner Francis Crick (co-discoverer of DNA together with James Watson[1]) terms this 'The Astonishing Hypothesis'.

> The Astonishing Hypothesis is that 'You', your joys and your sorrows, your memories and your ambitions, your sense of identity and free will, are in fact no more than the behavior of a vast assembly of nerve cells and their associated molecules. As Lewis Carroll's Alice might have phrased it: 'You're nothing but a pack of neurons (Crick 1994).

Marvin Minsky, a researcher of artificial intelligence at MIT, calls this view of the self 'The Society of Mind' (Minsky 1988). And if you think about it, it is hard to envision that the mind could work in another way. We *know* that the brain consists of a multitude of small branching units turning on and off, turning other

[1]The Nobel Prize in Physiology or Medicine 1962 was awarded jointly to Francis Harry Compton Crick, James Dewey Watson and Maurice Hugh Frederick Wilkins 'for their discoveries concerning the molecular structure of nucleic acids and its significance for information transfer in living material'.

small branching units on and off. We also *know* that the experience of a self is much too complicated for it to be explained through a single executive neuron.

Our selves *must* therefore at their innermost be built from many such tiny units that turn on or off. As simple as that—it just *seems* difficult. It may *seem* as if some internal unit is calling the shots—but that is just an illusion, *an illusion of self*. It is a very strong illusion too, so strong that you probably didn't have a problem with the word 'illusion'. Who is supposed to be having this illusion? To have an illusion necessitates a self, doesn't it?

Is everything—including our notion of self—an illusion? Yes, according to philosopher Daniel Dennett. He calls the sense of self a 'benign user illusion' (Dennett 1992). This does *not* mean that the self—consciousness—doesn't exist, merely that the self is not what it seems to be. So what is it?

Surely There Is Something More?

Some researchers and philosophers don't agree that this is how the mind works. Some claim that deep inside there must be a unit that 'understands'—and that without such a unit we have no explanation for understanding.

Prominent among them is the American philosopher John Searle who has tried to kill the idea that consciousness is an emergent phenomenon from a collection of cooperating neurons by invoking a thought experiment called 'The Chinese Room' (Searle 1980).

Imagine a computer programmer has managed to construct a computer that can communicate in Chinese. A person knowing Chinese can enter sentences using a string of Chinese characters and receive a string of Chinese characters in return. The program is skillful enough to pass the so-called Turing test (Turing 1950): a person using it cannot differentiate between conversing with a real Chinese person from conversing with the computer.

Question is, do the computer and program understand Chinese? There seems to be no unit inside the computer that understands Chinese, so can you *really* claim that the unit understands Chinese just because it does everything that a Chinese speaking person can do?

Searle answers no, and provides the following reason. Picture yourself in a room with access to all the algorithms that the computer you previously imagined used to communicate in Chinese. You have full access to the computer program. A piece of paper with Chinese characters pokes in through a hole in the door.

Your task is to do what the computer just did: follow the instructions of the computer program and produce a string of Chinese characters as an answer. These you feed through the hole in the door and thus produce the same answer as the computer and the program produced in the previous example. The Chinese person trying to figure out if you are a computer or another Chinese speaking person again cannot say.

But this time we *know* that the Chinese speaker is making a mistake—you *don't* know Chinese (if you happen to, please substitute language). There is thus no unit present in the Chinese room that understands Chinese. Therefore, Searle concludes, an understanding brain cannot be built from non-understanding units.

Searle's thought experiment has resulted in an enormous number of comments, mostly pointing out errors. I will focus on a single one of these answers: 'the system response', that it is the system's performance that is what counts.

The solution to the dilemma is—unsurprisingly—cooperation. Even if you don't understand Chinese, and the program doesn't understand Chinese, you and the program *together* understand Chinese. If comprehension turning a string of incoming Chinese characters into an outgoing string of Chinese characters, then understanding emerges from the interaction between you and the program.

What else is comprehension but this? Is there any other way to determine understanding besides looking at out-going information as a response to in-going information? Searle is very dismissive of the system response but shouldn't be. Almost all nature is made up of smaller parts—life itself!—so it is unsurprising that this holds true also in this case. Even if there would be a part of the brain that understands Chinese, this part itself is made up of smaller, non-comprehending parts.

Some people intensely dislike the idea of the society of mind because of the point of view. All we *really* know is what we experience. This must mean that we should start at that end? They insist that to use 'the material' to explain experience when all we have access to are experiences is to start at the wrong end. And since we experience that we are a unit, we are.

But can we trust experience enough to base science on it? One immediate problem with starting with experiences is that we each have direct contact with only our own experiences, not those of others. There is no guarantee that I am experiencing the same things that you are.

It is a reasonable assumption that it hurts equally when your hammer hits your thumb as when I hit mine, even though I do not experience any pain when you hit your thumb, only when I hit my own. Experiences are likely to be comparable. However, with this deduction I have already gone beyond my own experience to make statements about what is *not* my experience.

Basing research on own psyches and experiences in this way locks the researcher inside her own head, something that limits the scope of science incredibly, if taken seriously. The moment you want to step outside your own brain, all information has to be filtered through sensory organs. And to make inferences about things outside the brain you have to use scientific methodology, methods that are observer independent, outside experience.

Some researchers have other problems with the materialistic, soul-less description of your 'self'. Nerves reacting to nerve signals that react to external signals imply a very deterministic system, something that—it is argued—leaves no room for free will and thus undermines social responsibility. They don't *want* a world without souls.

Some researchers insert 'something more' to allow us the possibility to rule ourselves and take back control over our lives. Surely we determine *ourselves* how to act in the world—this we *know* because of our own experiences. Or is free will also an illusion? Well, we don't really know. But to base scientific theories on wishful thinking is never a plan.

A British mathematical physicist named Roger Penrose has proposed that the 'something else' we are looking for may turn out to be quantum events in the cell's microtubules (the cell's 'skeletons') (Penrose 1989). This is where free will would reside. Quantum events are to a degree unpredictable, so this would free us from determinism.

But critics say he is just moving the problem one step inwards, to a smaller scale, inside the nerve cells. And even if we add random quantum physical process, the brain still just reacts to signals from itself and the environment. If we didn't have free will before we don't have it with Penrose's quantum explanation—random events are not chosen.

Put more harshly: quantum events in the microtubules provide no increased understanding of consciousness. As the Canadian-American philosopher Patricia Churchland (1998) says, 'Pixie dust in the synapses is about as explanatorily powerful as quantum coherence in the microtubules.' Quantum processes may very well be involved in the brain's functioning, that's not the point of Churchland's quip. The problem is rather that quantum processes are also physical processes and not magical 'events of free will'.

It seems that we will have to be content with consisting of 86 billon nerve cells with 860 trillion connections. How all these cooperating parts result in each of us experiencing a unified self is one of the great riddles of science.

Nevertheless, cooperation in the brain works in a similar way to the cooperation we have previously investigated. The actions of many similar units (neurons) that are organized in specialized groups (parts of the brain) result in a unit with a function that is larger than the sum of its parts.

The existence of our mental selves is one of the great triumphs of cooperation. To build a coherent whole from small parts where the whole actually believes it is a unit—to build a unit that *believes*—is truly remarkable.

But is this riddle really any larger than the mystery that our bodies are units despite being built from many billions of cells? Even though we have known since our schooldays that we are built from cells, most of us don't walk around with a disorienting feeling of disunity.

Isn't the psychological self different, though? Well yes, we are missing something crucial. We can't understand the brain solely by understanding the physical processes of the brain. This is like trying to understand music through the physical workings of a stereo system. We have managed to disassemble a TV into its parts but we haven't considered Oprah. We have described the hardware but we haven't yet touched on the software.

We will return to the software in the two final chapters. Before this we have more steps to take. It is high time to contemplate cooperation between multi-cellular individuals equipped with the miraculous executive organ we have just considered, the brain. Does possessing a brain make any difference to how cooperation works?

To answer this we must first look at the question from the other direction. Is it possible to explain cooperation in animals that have a brain, without invoking the brain in the explanation?

References

Azevedo, F. A. C., Carvalho, L. R. B., Grinberg, L. T., Farfel, J. M., Ferretti, R. E. L., Leite, R. E. P., et al. (2009). Equal numbers of neuronal and nonneuronal cells make the human brain an isometrically scaled-up primate brain. *Journal of Comparative Neurology, 513*, 532–541.

Churchland, P. (1998). Brainshy: Nonneural theories of conscious experience. In S.R. Hameroff, A.W. Kaszniak, & E.C. Scott (Eds.), *Toward a science of consciousness: The second Tuscan discussions and Debates* (pp. 109–126). MIT Press.

Crick, F. H. C. (1994). *The astonishing hypothesis: The scientific search for the soul*. Scribner reprint edition.

Deacon, T. W. (1997). *The symbolic species: The co-evolution of language and the human brain*. Allen Lane: The Penguin Press.

Dennett, D. C. (1992). *Consciousness explained*. London: Penguin Books.

Edelman, G. (1987). *Neural darwinism: The theory of neuronal group selection*. New York: Basic Books.

Minsky, M. (1988). *Society of mind*. New York: Simon and Shuster.

Penrose, R. (1989). *The Emperor's new mind: Concerning computers, minds, and the laws of physics*. Oxford: Oxford University Press.

Rowling, J. K. (2007). *Harry Potter and the deathly hallows*. London: Bloomsbury

Searle, J. (1980). Minds, brains and programs. *Behavioral and Brain Sciences, 3*, 417–457.

Turing, A. (1950). Computing machinery and intelligence. *Mind LIX, 236*, 433–460.

Easily Explicable Cooperation and Natural Selection

Abstract Cooperation when there is mutual benefit is easily understood, but other cooperation is notoriously hard to explain. To fully appreciate the problem one needs a good understanding of natural selection, so a short summary is provided. The problems with group selection are summarized and behavioral genetics introduced.

It is time to look at cooperation between individuals. We now know that a multi-cellular individual is a conglomerate of conglomerates of genes, organelles, cells, organs and neurons that together form a unit that can move and interact with its surroundings. What about cooperation between such individuals? How do we get from individuals to societies? To understand cooperation between humans we have to start with the questions of how and why animals cooperate. We are, after all, a kind of animal.

The belling of a male red deer—a stag—can under beneficial circumstances be heard more than five kilometers away. This call is a demonstration of strength aimed at other deer. During rutting, each stag attempts to attract as many hinds as possible, while the hinds shop around for the most impressive stags.

Sometimes a competing stag will challenge the belling stag, attempting to impress and entice the hinds. Most often, it is enough for the harem holder to display, bell and maybe charge to convince the challenger to leave. But sometimes the challenger is a worthy competitor, someone for whom a mere demonstration of strength does not suffice.

In such cases, further comparison of size ensues. The stags measure each other up by belling. Then they walk besides each other to compare size. If this still does not settle the issue, an aggressive interaction is next. A clash of antlers follows and the two stags push each other with increasing degrees of violence. Stags are even killed under these circumstances (Clutton-Brock and Albon 1979).

Many fights between animals follow the same pattern. The battle does not start immediately, just from the two rivals seeing each other, but after a long check-list of behaviors where both individuals determine each other's size and strength. Either of the parties can withdraw at any stage before the violent fighting begins.

P. Lindenfors, *For Whose Benefit?*, DOI 10.1007/978-3-319-50874-0_4

Why do animals behave like this? For the weaker competitor the equation is perhaps not so hard to work out. The most probable result of entering a violent fight with a superior competitor is death or injury, not victory. Discretion is often the better part of valor. But the superior competitor also gains by avoiding violence and wasting energy on an obviously weaker competitor with the added risk of injury.

In this way many animals invoke a strict form of cooperation between combatants in a conflict situation. Both animals go through a pre-determined—genetically inherited—sequence of behaviors that progressively reveals the real strength of the animals, and thus minimizes the risk of bloody conflict. Two Swedish zoologists, Magnus Enquist and Olof Leimar, have calculated the evolutionary benefits of these behaviors and shown that individuals acting differently end up worse—for example individuals who never fight or always fight (Enquist and Leimar 1983).

Even in the most competitive and violent of situations there is thus a degree of cooperation that can be explained by mutual benefits.

Another example of the interaction between cooperation and conflict is sex. In evolutionary terms, there is a cost of sex as a sexually reproducing individual only transmits half its genes, the other half being contributed by an individual of the other sex. When reproducing asexually, *all* genes are contributed to the next generation. Why do organisms take this cost?

A hint about what is going on comes from evidence from a study by evolutionary biologist George C. Williams, who observed species of lizards that are capable of reproducing both sexually and asexually. In these species, sexual reproduction is timed with periods of environmental variability. When conditions are more predictable, the lizards reproduce asexually.

One consequence of sex is that evolution occurs much faster due to recombination (the mixing of the parents' genomes)—in this way adaptations that have evolved in different lineages get combined. Thus, there is an evolutionary advantage in reproducing sexually when fast adaptations are necessary—such as in George C. Williams' lizards when the live in variable environments (Williams 1975).

The evolution of sex really demands a book of its own, if all details are to be included. For this book's purposes, it is sufficient to know that sex evolved very early in eukaryote evolution, about 1.2 billion years ago, and has then been inherited to all descendants from there: animals, plants, and fungi (Bernstein and Bernstein 2010). The evolutionary advantage of sexual reproduction seem to mainly be about novel and faster adaptations, so there are benefits for both reproducing parties. Particularly, it has been suggested that resistance to parasites has played a major role in maintaining sex. Parasites evolve too, so counter-adaptations are always necessary. When there aren't such advantages, sex may be lost—as has happened in the aforementioned lizards, as well as in many other organisms.

Sex is thus a cooperative problem that is explicable through mutual benefit for both participants. However, once sex has evolved, other problems ensue. One example is human pair bonding. In mammals in general, females are stuck with first carrying the offspring and then nursing them until they have a degree of independence. Males and females have a common interest in their offspring, but there is always the option for males to abandon responsibility and instead find another female to impregnate. How this plays out varies immensely across mammals.

In some species the presence of a male neither adds nor subtracts from the chances of the offspring surviving to adulthood. In these species, females take care of the offspring on their own. In other species, the offspring demand so much nurturing that the presence of two parents is crucial for success. Here, males stay behind and help rear the offspring. Where do humans end up on this continuum? Well, it clearly isn't in one of the extremes.

According to the *Ethnographic Atlas Codebook*, reporting data for 1231 cultures, polygyny—marriage between one man and several women—is a common occurrence in 48% of the world's societies. In another 37% polygyny occurs, if not to a large degree. Monogamy is the norm in only 15% of the world's societies. Polyandry—where one woman marries several men—has been recorded in a grand total of four societies, which corresponds to 0.3% of all recorded societies on earth (Gray 1998).

Had human marriage systems been designed from mutual cooperation to raise common children then *all* societies would have had monogamy as the norm. But this is not the way it is. On the other hand, human marriage systems are not shaped completely by male abandonment either. The current state of affairs reflects some sort of middle ground between these extremes. In sexually reproducing species, the two sexes can reach different solutions in how to best optimize reproductive output, a difference Darwin named *sexual selection*, which is a subcategory of natural selection.

Mutual Gain

By far the easiest form of cooperation between individuals to understand is where everyone acts in a way that is of benefit to all participants, as seems to be the case with sexual reproduction. If no individual pays any form of net cost and if everyone reaps benefits, then the explanation of cooperation is simple and self-evident: everyone benefits!

In this category we find one of the classic cases of egalitarian cooperation, the cooperation between fungi and algae in lichens. The two organisms making up lichens are so tightly symbiotic that they are on the borderline of counting as one organism in their right, not associations of organisms. Lichens can live in extremely harsh environments, growing just a few millimeters each year.

Though symbiotic, the two organisms of the lichens can sometimes be found living independently. For the fungi part of the lichen, this is uncommon and demands specific circumstances concerning nutrients, but for the algae it is more common because of their photosynthetic abilities.

To reproduce, lichens can fragment or form propagules that include samples of both organisms. But the fungi part of lichens can also form *ascomata*—reproductive structures that produce fungal spores. These spoors develop into lichens if they encounter the appropriate algae. Both the fungi and the algae benefits from this symbiosis and together they can colonize many different habitats that they can't colonize independently.

But just occurring together isn't necessarily a tell-tale sign of cooperation—individuals can group together for many reasons other than to cooperate. For example, bacteria that divide end up in a large aggregation because there are many nutrients in one place—they simply end up together in the place they originated without having chosen this place, or chosen to associate with other bacteria.

The same is true with certain plants that grow together because living conditions happen to be optimal in just that place. Why do blueberries form mats of many plants that grow together? Because the living conditions are good there and blueberries mostly spread through shoots. Note that blueberry plants cannot later choose to leave—they cannot select and reject cooperation in the same way as mobile organisms—animals—can.

But mobile animals can also just happen to be in one place without having chosen cooperation. Think of humans. Why do city dwellers on their way to work flock on roads and in commuter systems? All of us have a common problem—to get to work—but this does not necessarily mean that we want to cooperate with the people surrounding us. We cooperate in this situation only to the degree that we do not bump into, push or sit on our fellow human beings. But would not everything been easier and faster without other people there?

This type of flocking, where individuals have similar goals but where the end result for each of us would be better if the others hadn't been there, is competition more than cooperation. Bacteria, blueberries and commuters would all have more room and resources for themselves (and arrived at work more quickly) if they had been alone.

There is no clear boundary between these types of aggregations that emerge by themselves and flocks that individuals choose to join because they actually benefit from other individuals being there. One of the simplest proposals for this latter form of group formation was suggested by the British evolutionary biologist Bill Hamilton (whom we will get to know better in next chapter). Hamilton suggested a mechanism he termed 'The Selfish Herd' (Hamilton 1970).

The idea is extremely simple. Every individual that runs the risk of being caught by a predator has a greater chance of escape if it has another individual between itself and the predator. If everyone tries to follow this simple rule, the result will be an aggregation of animals.

Since the middle of the herd is the safest and the outside least wise, all individuals will gravitate towards the middle. Individuals with better competitive abilities will collect in the middle while weaker individuals will be pushed to the periphery. The larger the flock you are part of, the better chance each individual will have to avoid the outside position.

Large group size is actually another, related example of mutual gain—everyone may benefit from just belonging to a large herd, as this provides some protection against predators and may benefit foraging.

To use other individuals as protection largely explains the enormous herds of hoofed animals on the African savannahs. Here you don't even have to cooperate with your co-specifics in order to reap the benefits of herd life—you can gain the same advantage from associating with individuals of other species. Zebras particularly seem to benefit by tagging along with the gigantic wildebeest-migrations across the savannahs (Sinclair 1985).

In a flock, there is the opportunity of gaining advantages from others in other ways. For example, more individuals mean more eyes to guard against predators. This leads to less pressure on individuals to stay constantly vigilant. As long as you look up every now and then with irregular intervals and others do the same, the herd will be more vigilant than any of the individuals on their own, without extra effort or organization (Dugatkin 1997).

Hunters can also benefit from working together. It is easier for several individuals to spot potential prey than for one. Through coordinated effort, the hunt can be more effective and larger prey can be killed than through individual effort. Lions benefit from their prides and will regularly take down animals that they never could have overpowered or killed by themselves. To live in a pride is a form of cooperation that is of mutual benefit for all individuals.

Finally, there is defense against predators. African buffalo stay tightly together when attacked. They place the calves in the middle and large, adult individuals on the outside. Buffalos will even charge predators together if another buffalo has been seized as prey.

But now we encounter a problem. Wouldn't it be beneficial for each buffalo to just hold back a little and let others do most of the defending? There is a potential cost to pay for the buffalo that ends up closest to the lion—the possibility of being wounded and maybe even killed. This risk can be minimized through not participating in risky actions. So why would any individual buffalo risk its life for others?

The riddle about cooperation emerges again at this point, when a cooperating individual seems not to gain from the cooperation. Why would any individual ever risk harm in the way buffalos do when they defend each other against lions? In evolutionary terms this should never happen—shouldn't all individuals prioritize their own survival? Surely, individuals who shy away from such risks must survive better and thus leave more copies of their genes in the next generation? To fully appreciate the solution to this problem we need a good understanding of how evolution works.

Natural Selection

Natural selection is a surprisingly simple process and at the same time remarkably powerful. Astoundingly, it was not understood until the middle of the 19th century. What Charles Darwin (and Alfred Russel Wallace) discovered was a creative process that doesn't require a creator (Darwin 1859).

The thought of a creation without a creator was hard to accept in some religious contexts. Especially as it was a popular belief at the time that the incredible ways that many organisms are adapted to their environment was one of the strongest arguments for God's existence.

For the same reason you can still find—after more than 150 years!—religious people who deny the existence of organic evolution on grounds of pure faith. It can therefore be worth looking closer at the theory of evolution and what you have to accomplish to prove it incorrect. As any scientific theory it is perfectly possible to disprove evolution with some valid observations. Scientific honor, fame and a Nobel Prize await the person who can.

The term *Evolution* is used in biology for two different phenomena. The first describes what has happened: that life has evolved from a common origin during millions of years to the diversity of organisms that we find today. In that sense, the theory of evolution is a description of a historical fact: that life has a history.

This part of the theory of evolution is extremely easy to prove wrong: just go out and excavate particular fossils! The history of life progresses according to a certain order; single cells came before multi-cellularity, a notochord came before the spine, and so on. The British evolutionary biologist J. B. S. Haldane supposedly once stated that he would renounce the theory of evolution if someone could discover a fossilized rabbit in Precambrian rocks[1] (this is more than 540 million years ago—long before mammals had emerged). It isn't harder than that to falsify this aspect of the theory of evolution; you just have to find the right fossil. But this demands dedicated work, and is not something that can be done in an armchair with a bible.

That life has a long history was no big surprise for Darwin's scientific colleagues. The evidence that the earth is very old had been accumulating for a long time. Fossils were present in collections around the world, showing long since extinct animals with a clear relatedness to present-day animals, yet radically different. Organisms showed strange likeness where those living in closer proximity often being more alike than animals living far away but in similar circumstances. Organisms had similar basic bauplans, revealing modifications around a theme.

The pattern was present for all to see: all life is related. What was missing was a mechanism for how so many life forms could emerge from a common origin. This was the crucial problem that Darwin solved—this was his real accomplishment. And here we have the second use of the term evolution, as a synonym to 'natural selection', the mechanism according to which life has evolved, how it happened.

---

[1]Possibly apocryphal. Even so, it illustrates well Haldanes' personality (he was known for his pithy comments) and a method that could be used to disprove the theory of evolution.

Natural selection can be broken down into three observations and one deduction. To show that natural selection is wrong you therefore have to show either that the one of the observations is false, that the deduction does not follow from the observations, or that the process isn't powerful enough to be the cause of the diversity of life.

For natural selection three things are needed:

- Variation—individuals of the same species differ.
- Inheritance—some of what determines how an organism turns out depends on inheritance from its parents.
- Unequal reproduction—all individuals do not have the same number of surviving offspring.

If you think about it, there is ample evidence that these three ingredients hold true. For humans, variation is obvious. Just look around in any crowd and you will find astonishing diversity. Then contemplate the fact that humans are specialists in finding differences among humans. Other organisms show the same variation as we do, even if we don't see it as clearly. There exists a simply astonishing diversity among all organisms, this we can observe.

Inheritance is equally obvious. A blue-eyed person almost always has two blue-eyed parents (though a mutation may also be a cause of blue eyes). However, it is one of the great insights of modern evolutionary science that you can't view inheritance and environment as two separate factors that each have an influence over how the organism turns out. All traits in an organism are products of how genes are expressed in certain environments. Nevertheless, the important thing to understand at this point is simply this: you inherit traits from your parents (mainly through DNA—genetic inheritance). This we can observe.

That all individuals do not have the same number of surviving offspring is also a trivial observation. Everywhere you study life you can count the number of surviving offspring. This is also an observable fact.

Natural selection is what you get when you have all these three conditions and draw the logically necessary conclusion of what follows.

It is not random who gets the larger number of surviving offspring. One animal may be better at nurturing its offspring than another, which increases the likelihood that its offspring will survive. Another animal might be stronger and can out-compete others at a source of nutrients. A third may be smaller and since it needs less energy to survive will do better when resources are scarce.

There are many factors that can influence the number of surviving offspring. Traits like being a better nurturer, having strength and needing few resources are just three examples. And a part of what makes one individual better than another is determined through inheritance; due to differences encoded in the DNA.

The number of genes that result in beneficial traits will become more common in the next generation because of the unequal reproduction. If the same conditions apply also in next generation, the same will happen again and again and again. Slowly the population will change to contain more copies of the beneficial genes relative to other, competing genes than before. Over a long time, this can cause enormous changes in the construction of organisms. Note also that when *all* organisms have some favorable trait, this changes the conditions so that other traits determine who gets more offspring. It is a never-ending process.

Just as with the initial genetic replicators, it is possible on this level as well to determine what separates good traits from bad traits. Good traits (or actually, genes resulting in good traits) are those that will enable their bearers to have more offspring than the competition.

But just as important as to understand how natural selection works is to realize what natural selection does not do. Three things in particular need to be pointed out.

1. *No individual changes*. It is the population that changes, not the individual. In no case is an individual born with one genetically encoded trait, discovers that it would be better to have another trait, and changes. Instead, traits are fixed in individuals at conception. It is the population that changes through differential survival and reproduction of individuals; individuals never change their genes.

   Things that are *learnt* aren't transmitted via genes to the offspring but must be learnt by each individual independently. Also, there is a phenomenon termed 'epigenetics' where experiences of a parent modify the expression of genes in the offspring. If a parent for example starves, this can influence the expression of genes for digestion in children. Epigenetics is the exception to the rule that experiences aren't inherited. Epigenetic effects can in this way result in quick evolutionary modifications of traits that last over a few generations.
2. *No new species emerges*. To produce new species more processes are necessary in addition to natural selection, for example that a population splits into two where selection acts on each half until they are different enough to not be able to have offspring together. If this happens, new species have evolved. Just understanding natural selection is not enough to understand speciation.
3. *No new trait emerges*. Natural selection acts on already existing traits. Traits do not *emerge* because of natural selection but through mutations and recombination (the latter process mixes parental genes at the time of sexual reproduction).

However, mutations are a rather misunderstood process. There has never been a mutation that from one day to the next turned an ape into a human. Who would that human mate with? Instead, many small changes over long time cause the transitions between forms. No female of any species has ever given birth to an individual of a different species.

Natural selection changes species slowly, similarly to the way growth occurs from childhood to adulthood. Almost invisibly, but irrepressibly, a child changes from day to day from a newborn baby to an adult individual. You can't see any difference from one day to the next, but looking at a photo of the newborn baby and comparing it with the adult individual, it is hard to see that it is the same individual.

Proximate and Ultimate Explanations

Here is a good place to clarify something that many non-biologists get hung up on—the question of motives. Take sex, for example. When you have sex, is it to get babies or because it feels good?

Most sexual intercourse taking place in the world has very little to do with reproduction. On the contrary, there are numerous methods, devices and pills to make it possible to make love without this resulting in a pregnancy. But despite the fact that this protection makes it evident that most sex occurs 'because it feels good', biologists claim that sex evolved because it results in babies. How does this compute?

It helps to understand the distinction between proximate (close, direct) and ultimate (definite, final) causes. Within philosophy, a proximate cause is the cause closest to the event that is to be explained. The higher-level, ultimate explanation is instead the 'real' explanation of what happens.

As an example, you can give two different answers to the question 'Why did the boat sink?'

- Proximate explanation: because the hole under the waterline made the boat fill with water.
- Ultimate explanation: because someone wanted to kill the passengers and therefore blew a hole in the hull.

In a murder investigation, the focus is nearly always on the ultimate cause of why someone died. Of course it is important that someone was stabbed with a knife or shot (proximate cause of death), but more important for the police is to find out if the perpetrator was the butler, who did it to gain an inheritance (ultimate cause of death).

In a similar manner, the ultimate cause is what is interesting to evolutionary biologists. On the question 'Why do people have sex?' the answer 'Because it feels good' (the proximate answer) is incomplete as an explanation, while 'To have offspring' (the ultimate answer) explains the evolutionary function of the behavior.

Often, the ultimate cause also explains the proximate cause. Sex feels good because that connection has resulted in more babies. That is why 'To have offspring' is the ultimate—evolutionary—cause of sexual intercourse. But it is also possible to go one step further and come even closer to the 'real' explanation. Why is having offspring important, really? Now we are back to the genes again.

Genes that provide instructions that make their carriers experience sex as pleasurable have left more copies over the millennia than genes for sexual intercourse that haven't been tied to pleasure. It is therefore almost certain that other animals also enjoy sex—why else would they do it? Genes providing instructions that would make sex uncomfortable or painful wouldn't become very successful as replicators.[2]

Group Selection

After this brief foray into natural selection we return to the question of cooperation. When we left off, the problem we were contemplating was when one of the participants in a cooperative interaction appeared to *not* gain any benefit. Why would any individual sacrifice him- or herself in the manner that buffaloes do when they defend each other against lions? In evolutionary terms this surely should not happen—everyone ought to save their own skin first, right? A gene instructing self-sacrificial behavior would just disappear because of natural selection, as we just saw.

For a long time the explanation for self-sacrificial behavior in nature was that individuals sacrificed themselves *for the good of the species*. If someone warns the herd and gets eaten, the end result is *good for the species*—the others make it, even if one individual dies. *Of course* such self-sacrificial behavior for the good of the species should be naturally selected.

Is this perhaps why some birds stay at the nest and help their parents raise another brood instead of saving energy for next year's reproduction? For the good of the species? Similarly, why don't all individuals in some primate societies reproduce when there is not enough food? Do some individuals sacrifice their own reproduction for the good of the species?

In 1962 a British zoologist named V. C. Wynne-Edwards published a book titled *Animal Dispersion in Relation to Social Behaviour* (Wynne-Edwards 1962) where he reviewed a large number of such self-sacrificial behaviors and suggested that all these behaviors were adaptations that, although detrimental for the individuals, were good for the group.

This argument, which Wynne-Edwards presented thoroughly, is superficially similar to the argument presented earlier, in the discussion on multi-cellularity in the second chapter. It doesn't really matter if single cells die as long as the whole individual benefits. Similarly, according to Wynne-Edwards, it doesn't matter if some individuals die as long as the whole group benefits. Natural selection thus acts on the group, not on the individual.

---

[2]This is not universally true. There are for example species of bugs where sexual intercourse consists of the male boring a hole through the female's exoskeleton. This has evolved because it circumvents female choice, so that males having this adaptation have had more offspring than males that didn't have it. This is most certainly not true for the females, however, so any female that by chance happens to get a mutation for more efficient armor would probably have a higher reproductive output than her conspecifics. This mutation just hasn't happened yet, in those species.

Within the field of evolutionary biology, V. C. Wynne-Edwards is somewhat unfairly remembered mainly for this faulty reasoning. When he wrote his book, group selection was widely thought to be correct. What Wynne Edwards did was to review the evidence so thoroughly that, ironically, it became apparent that the explanation couldn't work. It was principally another evolutionary biologist, the American George C. Williams, who demolished Wynne-Edwards and other group selectionists' explanations (Williams 1966).

The problem that Williams pointed out is not so difficult to see once it is highlighted. Imagine for a moment that the age-old tales of lemmings are true (they're not) and that lemmings actually migrate to the coast to collectively drown themselves. They can't *all* do this, of course, because then there wouldn't be any lemmings left to produce future generations, so just imagine that *many* wander off—for the good of the species.

The problem with this story is that in principle *all* biological traits have a genetic component. Now ask yourself, what type of genes become most common in the next generation: genes that provide lemmings with an insatiable lust to drown themselves or genes that make the lemming prefer to stay and reproduce?

Well, *of course* there will be more copies of genes making the lemming stay and reproduce in the next generation, how could it be different? And the same applies to *all* self-sacrificial behaviors. Genes that result in self-sacrificial behavior *can't* become more common through natural selection—the very act of self-sacrifice will weed such genes out of the population.

Note the difference with multicellularity. Multicellular organisms consist of cells with the *same* genes (fraternal cooperation) while lemming groups consist of individuals with *different* genes (egalitarian cooperation).

Cells in a multicellular organism can un-problematically sacrifice themselves for the group, because their common genes are transmitted through a unicellular stage, and it is this transmission that makes the difference—the transmission that carries the genetic package on to the next generation; not the survival of individual cells within the multicellular organism. Or, to phrase it differently, it doesn't matter for your reproduction if you happen to chop off your finger.

Individuals in lemming groups cannot do the same thing, because the copying of any one individual's genes is all down to the reproduction of that individual. Or, to phrase it differently, it matters for the individual's reproduction if it drowns.

While cells in a multicellular body have a common genetic interest, individuals in a lemming group don't (unless the circumstances are right—we'll get to that).

Note that this is the description of what was classically termed group selection. Group selection has later also come to be used to describe the observation that natural selection can act on many levels (gene, individual, group), something that is better termed 'multilevel selection', and the observation groups sometimes compete (West et al. 2011).

Multilevel selection is a useful tool to understand natural selection, the fact that adaptations for a well-functioning group are important for individuals who are included in that group. But this is not the same thing as trying to understand self-sacrificial cooperation in individuals by arguing that they sacrifice themselves 'for the good of the group'.

Behavioral Genetics

But hold on a second: genes and behavior? Surely behaviors are the result of thought processes in the brain, not of genes? It is one thing to make statements on genetic influence on body size, antlers and other physical traits, but behavior, really? We have already introduced brains in the chapter on the psychological self, and on this level of cooperation, surely they must count for something? We can't still be talking about genes?

Yet here we are. We still have to focus on genes. Brains are to a degree 'instructed' by genes. Behavioral genetics—the area of inquiry of the link between genes and behavior—is a research field, not some wacko idea. Behaviors are actually just as genetically influenced as all other traits.

Look at cuckoos, for instance. Cuckoos hatch from eggs that their mothers have laid in other species' nests. A cuckoo does not meet another cuckoo until it is time to mate. Everything 'cuckoo-like' in its behavior thus has to be innate. Nevertheless, as newly hatched, cuckoos react to the presence of other eggs in the nest and push them out—this is a reflex of being pressurized on the back, something that makes the chicks push backwards. Not only this, but the hatched cuckoos 'know' how to beg for food and how long to stay in the nest. No other cuckoo can have taught the baby cuckoos all these things—cuckoo chicks have never met another cuckoo, ever.

If the cuckoo is a male, he will also know how cuckoos sing as well as where and when to do so. He knows how close his territory should be in relation to other males and what type of males he should avoid. If the cuckoo is female she will know in what birds' nests she can lay her eggs and what nests to avoid. Cuckoos 'know' all this without having had the opportunity to learn it. Genetically encoded structures pre-existing in the cuckoos' brains make it happen.

Another example is horse ants—*Formica rufa.* Their nests are started by a single newly mated queen who starts off by laying a few eggs. These then hatch and worker ants emerge. These ants 'know' how to build an ants' nest, how to care for eggs, how to defend themselves against other ants, how to find food, when to build and when to forage, and so on. These first worker ants have never met other worker ants in their lives and thus haven't had any opportunity to learn from someone else. The queen teaches them nothing—they already 'know' what to do.

Some dogs are good at tracking, others at fetching, others at herding, yet others at guarding. Some dogs are aggressive and dangerous, others are friendly and sociable. These are traits that humans have looked for in dog breeding for many hundreds of years, using the knowledge that aggressive dogs beget aggressive offspring and friendly and sociable dogs beget friendly and sociable offspring. It is perfectly possible to breed forward behavioral traits; these are no different in this regard from other traits.

Eels migrate halfway around the globe to mate in the Sargasso Sea—without ever having made the trip before they can find their way there. Chickens having lived their whole lives in barren cages still perform the movement wild fowl carry out when they sand bathe—without ever having encountered sand. Male stags compete, female hinds do not.

What about cooperation? Well yes, wolves, chipmunks and chimpanzees live and cooperate in flocks while tigers, bears and orangutans live solitarily. Gibbons and seagulls most often form breeding pairs, red deer and gorillas most often form harem-like groups. Ants and bees cooperate in large societies, dung beetles and parasite wasps live alone.

So of course behaviors have a genetic component. This does not mean that we are robots, completely remote-controlled by our genes, just that there is an aspect of every behavior that is genetically influenced. In the same way as there are aspects of your height and weight that are genetically determined, the same is true of your behaviors.

These genetic aspects are why self-sacrificial group selection cannot work in nature. An individual with a gene that results in a self-sacrificial behavior pays a cost that other individuals don't have to pay. In the next generation those genes will therefore become rarer and in the next generation rarer still. Soon the gene will disappear and with it the behavior.

But can you conceive of cases where the benefit for the group is so great that it overwhelms the negative aspects of the behavior? Surely a group with self-sacrificial individuals can gain advantages compared with a group of egoists? As the great American biologist EO Wilson has stated in an interview: 'Competing is intense among humans, and within a group, selfish individuals always win. But in contests between groups, groups of altruists always beat groups of selfish individuals' (Lundborg 2012).

That is definitely true, but currently beside the point. The problem with animal groups is that individuals have much shorter generation times than groups and that groups aren't stable across generations (individuals move between groups). These two factors together mean that group selection will be a much weaker force than selection on individuals. To explain why some individuals sacrifice themselves for others we need a force that is *stronger* than natural selection on individuals, not weaker.

However, as we will see in a later chapter, group selection can work in a cultural context, i.e. in humans. And if you think about the earlier example of fraternal cooperation, cells in multicellular species, these die for the good of the organism. In this example you can find a hint to the solution provided in the next chapter, as all cells in a multicellular species share genes. In egalitarian cooperation, there is no comparable self-sacrifice—no organelle sacrifices itself for the good of the cell, no bacterium for the good of the host organism, no fungi or algae for the good of the lichen. Group selection, as it was classically formulated, cannot explain self-sacrificial behavior in egalitarian cooperation.

This means that we run into a problem, however. We have just dismissed an explanation of self-sacrificial cooperative behaviors without providing a better explanation. We still have to explain why buffalos risk their own lives to protect the herd against lions, why some birds sacrifice their own reproduction to work hard to help their parents—and what about ants and bees who make the ultimate sacrifice: they refrain from reproducing completely and instead devote their lives to helping their mother, the queen.

How can such behaviors be favored by natural selection? This question dogged Darwin who was never able to solve it. But he was on the right track:

> This difficulty, though appearing insuperable, is lessened, or, as I believe, disappears, when it is remembered that selection may be applied to the family, as well as to the individual, and may thus gain the desired end.
>
> Charles Darwin: On the Origin of Species (1859).

References

Bernstein, H., & Bernstein, C. (2010). Evolutionary origin of recombination during meiosis. *BioScience, 60*, 498–505.

Clutton-Brock, T. H., & Albon, S. D. (1979). The roaring of red deer and the evolution of honest advertisement. *Behaviour, 69*, 146–170.

Darwin, C. (1859). *On the origin of species by means of natural selection, or the preservation of favoured races in the struggle for life* (1st ed.). London: John Murray.

Dugatkin, L. A. (1997). *Cooperation among animals: an evolutionary perspective*. Oxford: Oxford Series in Ecology and Evolution.

Enquist, M., & Leimar, O. (1983). Evolution of fighting behavior: decision rules and assessment of relative strength. *Journal of Theoretical Biology, 102*, 387–410.

Gray, J. P. (1998). Ethnographic Atlas codebook. *World Cultures, 10*, 86–136.

Hamilton, W. D. (1970). *Geometry for the selfish herd*. Dissertation, Imperial College.

Lundborg, Z. (2012). Wilson says greedy people, altruistic groups win. *Bloomberg Careers*, 17 April.

Sinclair, A. R. E. (1985). Does interspecific competition or predation shape the African ungulate community? *Journal of Animal Ecology, 54*, 899–918.

West, S. A., El Mouden, C., & Gardner, A. (2011). Sixteen common misconceptions about the evolution of cooperation in humans. *Evolution and Human Behavior, 32*, 231–262.

Williams, G. C. (1966). *Adaptation and natural selection*. Princeton: Princeton University Press.

Williams, G. C. (1975). *Sex and evolution*. Princeton: Princeton University Press.

Wynne-Edwards, V. C. (1962). *Animal dispersion in relation to social behavior*. London: Oliver & Boyd.

Family

Abstract Hamilton's work resulting in Hamilton's rule is presented. Some classic examples of how this explains seemingly self-sacrificial behaviors are presented. Amongst these are cooperation in ants and bees, but also in termites and naked mole rats—and humans. I discuss the current controversy initiated by work of EO Wilson and Martin Nowak.

When reasoning about costs and benefits in evolutionary biology it is always reproductive success that is the measure, or more formally correct, the entity termed *fitness*: the difference in reproductive success of an individual, gene or trait in relation to another individual, gene or trait. To formulate the problem of self-sacrifice in these terms is to ask the question when it is beneficial for an individual to sacrifice a part of its own reproductive success to benefit another individual's reproductive success. Shouldn't such behaviors always result in lowered fitness?

If we look at the situation from an evolutionary view-point, however, the question isn't best phrased why certain *individuals* sacrifice a part of their reproductive success, but rather why certain *genes* coding for self-sacrificial behaviors become more common in a population. This distinction may seem hair-splitting—what does it matter if you count individuals with a trait or the number of individuals having genes encoding that trait—isn't it the same thing? As we will see, it makes a world of difference.

The new formulation of the problem of self-sacrifice thus becomes: when is it beneficial for a gene to instruct an organism to sacrifice some of its reproductive success to benefit another individual's reproductive success?

This shift of perspective is one that Darwin never achieved. How could he? Darwin knew nothing about genes. Self-sacrificial behavior was thus extremely difficult for him to explain, and for those who came after. It would fall to the aforementioned evolutionary biologist Bill Hamilton to solve the problem.

In the beginning of the 1960s Hamilton was a Ph.D. student in England. He had long tried to study evolution and self-sacrificial—or altruistic (truly unselfish)—behavior in nature, but no-one believed in his ideas. He kept on stubbornly

P. Lindenfors, *For Whose Benefit?*, DOI 10.1007/978-3-319-50874-0_5

working, however, sometimes believing that he was a genius alone with the truth, sometimes wondering if he was working with obvious truths or obvious falsehoods. After all, most people who believe they are on to something truly revolutionary are wrong. Hamilton even considered abandoning science and becoming a carpenter instead.

Work progressed slowly. In his autobiography he writes that his problem was his insufficiency in mathematics, but he needed mathematical calculations to solve the problem. He had to teach himself what he needed to know.

Each time Hamilton thought he had arrived at the solution, he found yet another error. He felt so alone with his quest, he would sometimes travel to Waterloo station and sit there reading or doing his calculations, just to be surrounded by people.

Hamilton felt that his chances of getting a Ph.D. were slipping through his fingers. To increase his chances of getting hired anywhere, he threw together a summary of his findings in 1963 in an article titled *The genetical evolution of social behaviour*. After the article had been accepted and was to be published in *The Journal of Theoretical Biology* the journal asked Hamilton to split the paper into two parts—one theoretical, more mathematical half and one half with a verbal description of the theory, and examples (Hamilton 1964).

These two articles are the most cited articles of the *Journal of Theoretical Biology* ever and are now considered fundamental classics in evolutionary biology. The articles led to books such as George C. Williams' *Adaptation and Natural Selection* (Williams 1966), that demolished group selection, but also to Richard Dawkins' *The Selfish Gene* (Dawkins 1976).

Hamilton's articles are (remarkably, considering their influence) seldom read. In particular, the mathematical first part is practically impenetrable for a non-mathematician. That they are seldom read has been revealed through analyses of references to the articles where errors in citations have spread through the scientific literature in a way indicating that researchers have copied the citations from other articles instead of citing the articles themselves.[1]

Not having read the articles does not mean not understanding their content, however. Hamilton's insight—the finding now termed Hamilton's rule—is basically very simple, so simple that you don't need to read the original articles to understand it. (This is probably lucky for you, dear reader.)

To fully understand Hamilton's results you first need to know something about coefficients of relatedness. Such a coefficient is *the probability that a specific gene on a specific place in the genome is identical in two individuals because they both have inherited it from a common ancestor*.

Think of it this way. You have inherited half of your mother's genes. So has your full sibling. But you haven't inherited the same half. Genes are shuffled (recombined) before sperm and eggs are constructed. This shuffling results in your sibling inheriting about half of the half you yourself inherited from your mother.

---

[1]The citation was to 'The Genetical *Theory* of Social Behaviour' instead of the correct 'The Genetical *Evolution* of Social Behaviour'. Whatever that proves....

Exactly the same applies to the inheritance from your father (except that if you are a boy you got his whole Y-chromosome and if you are a girl, his whole X-chromosome—in both cases you also get an X-chromosome from your mother).

This means that one fourth of your genes will exist in identical copies in you and your sibling because you both got them from your mother, and one fourth will be identical because you both inherited them from your father. Thus, full siblings have the coefficients of relatedness $r = \frac{1}{4} + \frac{1}{4} = \frac{1}{2}$. Biological full siblings thus have half of their genes in identical copies because of common descent.

For a randomly selected gene you are carrying, the probability is thus 50 percent that your full sibling also has it because you both got the same copy from your parents. You can see the calculations of a few coefficients of relatedness in Fig. 1. Note that this is a somewhat simplified explanation of r—more formally, r measures genetic similarity between socially interacting individuals relative to the rest of the population (Hamilton 1970; Grafen 1985). This difference will matter later.

In reality, all people have more than 99% of the base pairs in their DNA in identical copies (Jorde and Wooding 2004). A large part of the active sections of the human genome consists of instructions for how the cell works, other genes are common for mammals, yet others for apes and so on, all the way up to the approximately 99% of the genes that we have in common with chimpanzees (Jorde and Wooding 2004). These are not the genes that make it so crucial to understand coefficients of relatedness.

Instead it is all about variable genes that come in multiple versions; those subjected to natural selection that can either become more common or rarer. If you have inherited one of these gene-variants, what is the probability that your full sibling also has that same gene-variant? Fifty percent. This is when the coefficient of relatedness becomes important.

Hamilton imagined a 'gene for altruism' (selflessness). If such a gene makes an individual sacrifice part of its reproduction for another individual, how big is the probability that the individual actually helps another individual carrying the exact same gene for self-sacrifice? This probability is the same number as the coefficient of relatedness.

If the gene for unselfishness can 'help itself' in other individuals, then the gene should become more common even if it makes individuals sacrifice their own

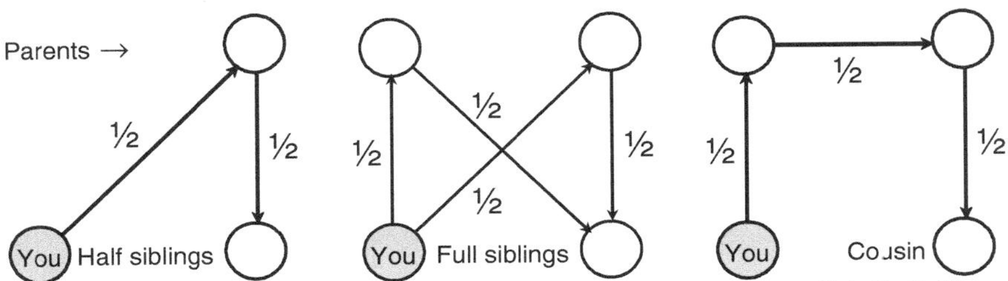

Half siblings: $r = \frac{1}{4}$ $(= \frac{1}{2} \cdot \frac{1}{2})$ Full siblings: $r = \frac{1}{2}$ $(= \frac{1}{4} + \frac{1}{4})$ Cousins: $r = \frac{1}{8}$ $(= \frac{1}{2} \cdot \frac{1}{2} \cdot \frac{1}{2})$

Fig. 1 Coefficients of relatedness—'you' are the gray circle

reproduction for other individuals. Hamilton summed up this insight in something termed Hamilton's rule (which is the closest thing biology has to Einstein's $e = mc^2$):

$$r > C/B$$

This is to be read that the coefficient of relatedness r has to be larger than the cost C (for Cost) divided by the benefit B (for Benefit). If these conditions are met, then a gene for altruism can become more common in a population because it helps itself, but in other individuals. The costs and benefits are all the time to be calculated *for the genes*, not for individuals. Evolution is all about changes in gene frequencies. Individuals die, remember, but genes are potentially immortal.

Simply put, Hamilton's rule states that the gene for altruism can spread if it benefits itself in other individuals. The gene must therefore code for behavior such that the carrier individual acts altruistically towards its *relatives*, and more so towards close relatives than distant. The British evolutionary biologist and geneticist J. B. S. Haldane is said to have once formulated the same insight in a pub discussion when he said that he wouldn't be willing to sacrifice his life for a drowning brother, although he would 'for two brothers or eight cousins'.

Hamilton's formula is the solution to Darwin's problem. With Hamilton's solution there is all of a sudden a mechanism that can explain why some individuals sacrifice their own reproductive output for others. Behaviors that at an individual level seem unselfish are at the genetic level more like genes selfishly helping themselves, independently of what individual they happen to be in.

It is from this insight of how selection plays out at the genetic level that Richard Dawkins got his metaphor of the selfish gene (Dawkins 1976). The 'selfish gene' of course doesn't imply that genes actually have a conscious purpose and consciously behave selfishly. But it is much easier to understand what is going on if you look at what is happening at the genetic level *as if* genes are behaving selfishly; this is the point of the metaphor.

Note, however, that this 'selfishness' applies to *all copies* of a gene—they are cooperatively selfish in that it doesn't matter if copy 1, copy 2 or copy 3 is copied, as they are all identical. When we speak of selfishness between individuals, we normally think of the fate of only one individual (West et al. 2011). For genes, the focus is on the information content, not the actual physical copy.

Think of a gene *as if* it is selfishly directing cooperative interactions between individuals so that they behave altruistically towards individuals carrying the same genes, then you may understand what Dawkins means. This is a metaphor that has helped thousands of undergraduates understand the intricacies of evolutionary biology.

The real irony is that many people—mostly those who have never read Dawkins' books or any evolutionary biology—believe that selfish genes have to result in selfish individuals. As we now can see, the opposite is true; the metaphor of the selfish gene has its roots in the evolutionary explanation of *unselfishness* and *altruism*. Think about that the next time that you hear someone ramble on about how

'Dawkins states that all organisms are selfish' (though to be fair, he does make a few remarks like that in *The Selfish Gene*).

So yes, genes can be seen as selfish—we know this through our understanding of cooperation in nature.

A gene for cooperation, or a gene for altruism, is not an abstraction, a metaphor or a figure of thought. As we saw earlier, behaviors—also complex behaviors—are often linked to genes. It is not by chance that flies live by themselves while ants cooperate in thousands, these are behaviors that are totally genetically determined.

But wouldn't it be just as beneficial for a gene to instruct its bearers to cooperate with everyone? Then it would still also be helping itself in other individuals? The answer to this question is no. If you give up your reproduction for everyone then all genes will be benefited equally. The only thing that happens is that the gene will pay the reproductive cost of self-sacrifice without gaining any benefit relative to others. For a gene of unselfishness to spread, individuals must sacrifice themselves *specifically for individuals carrying the same gene*—otherwise there will be no special advantage for the gene—no selection making it more numerous.

Note, though, that it works with a rule that individuals should cooperate with those close-by. The condition for this rule to apply is that there is limited dispersal. In cases where individuals do not move much, individuals in the vicinity will be family—related by common descent. Then Hamilton's rule applies, without kin recognition (Hamilton 1964).

Of course, no-one is stating that animals calculate coefficients of relatedness before behaving. Instead, genes are favored by natural selection if they result in behaviors in accordance with Hamilton's rule. Genes making individuals act according to the logic dictated by the coefficients of relatedness are naturally selected and become more common. Similarly, no animals have to carry out calculations including the gravitational constant to reach the correct speed when falling.

What Hamilton's rule explains is why relatives tend to cooperate with each other, and why there is more cooperation between close relatives than between distant relatives. This process is therefore termed *kin selection*, or *kin altruism*.

The methods genes use to coerce us to act in specific ways are *feelings*. Why do you love your children? (Most people do, though not all—variation exists in all biological traits). Why do people fall in love, become horny, angry, scared or happy? These are ancient systems to quickly guide us to act in a specific manner. Anyone who has ever been the victim of strong emotions knows what powerful language these instructions speak.

Warning Calls

A classic example illustrating how kin selection works in nature is the case of the warning calls of Belding ground squirrels (Sherman 1981). These animals warn each other upon spotting a predator. The problem is that the individual that cries out the warning runs a greater risk of being attacked, since the warning attracts the predator's attention.

The American biologist Paul Sherman, who has investigated this behavior, once saw a badger who had crept close, watching the squirrels. It remained unnoticed until one of the squirrels spotted it and directly proceeded to call out a warning. The badger immediately turned on the calling squirrel and killed it.

Death by predator is common in the ground squirrel environment; approximately 10 percent of the attacks Sherman recorded ended with the death of a squirrel. So why would any individual want to warn others when there is such a large risk of attracting the predator's attention and getting killed? To understand this you have to know a bit about ground squirrel biology.

Males in Belding ground squirrels disperse about ten times further away than females. Females thus live their whole life surrounded by their relatives, while males initially are surrounded by non-relatives. According to Hamilton's rule, we should thus expect females to warn to a greater extent than males.

If you investigate who calls, females do this much more frequently than males, who almost never warn. Females are warning their relatives who are in the same flock, while males who have moved to a new social context don't warn nearly as often.

The ability to recognize relatives is central for such a system of risky unselfishness to work. But if individuals could recognize each other perfectly, the coefficient of relation wouldn't have to be in Hamilton's formula—if the gene for altruism could signal its existence inside an individual it would be very easy to understand the spread of altruism through populations—individuals with the gene would only act unselfishly towards those carrying the same signal.

Richard Dawkins has popularized this idea by discussing a potential 'Green Beard' that is linked with cooperation. It would be possible to perfectly identify altruistic individuals—you can tell who carries the cooperative gene from their 'beard'. A few cases of such figurative 'green beards' have been found in nature, but most commonly there exists no such signal (Dawkins 1976). Note that the Green Beard effect is different from cooperation with similar types, though the latter also makes the evolution of altruism easier (Colman et al. 2012).

In a review 2010, biologists Stuart West and Andy Gardner noted that only five examples of 'Green Beards' have been found in nature—one in an ant and four in microbes. This is simply not a common path to cooperation, although it is theoretically possible (Gardner and West 2010). We will return to the Green Beard hypothesis later, in the chapter on languages.

Since there is no trait identifying bearers of the altruistic gene, better to cooperate only with your closest relatives. This is the reasoning behind the well-established system for recognizing relatives in Belding's ground squirrels.

Kin selection of the type that exists in ground squirrels is very common in nature (Dugatkin 1997). Most forms of costly cooperative behavior in animals are due to kin selection. Parents take great risks for their young—as with the African buffalos. Offspring go to great lengths to help their parents raise siblings—as with certain bird species where the young stay with their parents. Understand kin selection and you have understood most of animal cooperation.

Eusociality—Ants, Wasps and Bees

What about ants? Worker ants sacrifice *their whole reproduction* to help their mother the queen—how can such behavior be inherited when workers don't even reproduce? Let's take a look at this extreme form of self-sacrificial behavior, termed *eusociality*—'true sociality'—where just one or a few individuals reproduce while all others refrain and just help. How can it be advantageous to sacrifice your whole reproduction for another individual? And how is non-reproduction inherited anyway?

Or is this an incorrect description—would it be better to speak of a single 'super organism' where some parts of the organism reproduces and others do not, just as is the case with human cells (Queller and Strassmann 2009)?

Hamilton thought that a clue to the solution may be that this type of behavior is common in the insect order Hymenoptera (sawflies, wasps, bees and ants). The trait that some individuals sacrifice their whole reproduction has evolved eleven times in this insect order (Hughes et al. 2008) and just eight times in all other animal groups taken together. Why?

To understand Hamilton's suggested solution we have to return to the coefficients of relation; these are extremely different in Hymenoptera in comparison to those of humans because the genetic sex-determination is very odd in this group. We humans get half our genes from our mother and half from our father. If you happen to get a Y-chromosome from your father you become a boy, but if you instead get an X-chromosome you become a girl.

In Hymenoptera sex-determination occurs in another way completely. If an individual gets genes both from its mother and its father if becomes female. But if the mother lays unfertilized eggs—eggs that thus contain only half of her genes but no genes at all from any male—then these individuals will become male.

Males therefore only have half as many genes as females—half as many chromosomes—and all the males' genes come from their mother. This skews the coefficients of relation to be very different from those of humans, just look at Table 1.

From this table you can deduce that humans should help their close relatives to about the same degree (parents, children and full siblings). That we don't, but parents help their children more than children help their parents depends, biologically speaking, on the future *potential reproduction*. Young individuals have the

Table 1 Coefficients of relation for humans and ants

Humans	Daughter	Son	Mother	Father	Sister	Brother	Niece/Nephew
Female	½	½	½	½	½	½	¼
Male	½	½	½	½	½	½	¼
Ants	Daughter	Son	Mother	Father	Sister	Brother	Niece/nephew
Female	½	½	½	½	¾	¼	$^3/_8$
Male	1	0	1	0	½	½	¼

possibility of having offspring to a greater degree than older individuals. Genes enticing you to help your children are therefore more favored by natural selection than genes enticing you to help your parents (but don't worry about your old age just yet; humans are not only gene-controlled robots—there are more chapters in the book).

Now take a look at the relatedness coefficients of ants. First look at the coefficients for a female (follow the row with the row-heading 'Female' from left to right). A female who gets a daughter leaves half her genes to that daughter. But from the table you can also see how large a part of her genes that instead would be represented if she instead *helps her mother* to produce a daughter: three quarters (the coefficient of relation for sisters).

The reason for this skew is that half of a female's genes comes from her mother. But since the father only has half as many genes, he will leave *all* of his genes to his daughters. The end-result is that a worker-ant's genes—think especially about genes for altruism—will be more favored by natural selection if she helps her mother to raise sisters than if she herself becomes the queen and produces offspring of her own. Genetically, it is thus more advantageous to be a sterile worker helping your mother to produce new queens than to be a queen yourself.

Another thing you can read out of the table is that none of this is true for males, so you shouldn't expect any male to give up his reproduction for anyone. Males cannot have sons (since males hatch out of unfertilized eggs), but can have daughters, brothers and sisters. The relatedness to these is, however, equally large. There is therefore no special genetic motive for a male to help in the nest. This is why males—drones—live only to mate and then die. They don't help for one second. All worker ants or worker bees are females.

You can also read out other fascinating information from the table. For example, a queen has similar genetic motives of getting sons and daughters—the coefficients of relatedness for both are ½ since the queen in both cases supply half of her genes. But a worker ant has a coefficient of relatedness that is ¾ with her sisters but only ¼ towards her brothers.

With this we have identified a conflict of interest between the queen and the worker ants. This plays out in that the ant queen lays about as many unfertilized as fertilized eggs—she produces the same numbers of male as female eggs. But the worker ants want something else completely; they want the sex ratio skewed. They therefore take unfertilized eggs (males) and feed it to the larva from the fertilized eggs (females) until the sex ratio is in accordance with their coefficients of relation: ¾ sisters and ¼ brothers. The queens respond by trying to control the proportions by laying even more males. This evolutionary conflict has resulted in different sex ratios in different species of ant, all depending on if the queen or the workers are more in control for the moment (Passera et al. 2001)—animal class warfare.

The example of the social insects (ants, bees and some wasps) was Hamilton's *coup de grâce* toward the doubters who didn't want to relinquish this much importance to genes in explaining altruism in animals.

Bill Hamilton was a great insect lover through his whole life. Exactly *how* great was understood at his death (from an intracranial hemorrhage in March of 2000 due to repercussions of a severe case of malaria he came down with in the jungles of Congo). He had left detailed instructions for his burial in an essay entitled *My intended burial.*

> I will leave a sum in my last will for my body to be carried to Brazil and to these forests. It will be laid out in a manner secure against the possums and the vultures just as we make our chickens secure; and this great *Coprophanaeus* beetle will bury me. They will enter, will bury, will live on my flesh; and in the shape of their children and mine, I will escape death. No worm for me nor sordid fly, I will buzz in the dusk like a huge bumble bee. I will be many, buzz even as a swarm of motorbikes, be borne, body by flying body out into the Brazilian wilderness beneath the stars, lofted under those beautiful and un-fused elytra which we will all hold over our backs. So finally I too will shine like a violet ground beetle under a stone.

Despite these stated wishes he was never eaten by *Coprophanaeus* beetles. Instead, a secular burial was held for him at Oxford (Nowak and Highfield 2011).

But the story of cooperation in Hymenoptera has not turned out to be quite as clean as I have described it here. Problematically, the coefficients of relatedness are only those specified in Table 1 when all workers have the same father, and queens in some species of Hymenoptera mate many times. This means that it often is the case that workers are *not* more related to their sisters than they would be to their daughters. But this may be a secondary development, after eusociality has emerged, and thus an evolved form of cheating on a well-functioning system.

In other species it becomes more complicated still because multiple queens nest together. In these cases some workers have neither the same mother nor the same father. Again, however, this may be a secondary development, or something that has evolved secondarily through mutual benefit.

But as we shall see, skewed coefficients of relation are neither necessary nor sufficient for the evolution of eusociality. It takes very special ecological circumstances for the type extreme cooperation we find in ants, bees and wasps to emerge.

Moreover, haplodiploidy doesn't only lead to workers being more related to their sisters ($r = \frac{3}{4}$) but also to workers being *less* related to their brothers ($r = \frac{1}{4}$), something, it has been argued, that *exactly* cancels out the effects of the skewed relatedness on which Hamilton's theory for the evolution of eusociality rests. To make daughters you need both a queen and a drone, after all. And if workers are more related to their sons ($r = \frac{1}{2}$) than their brothers ($r = \frac{1}{4}$), wouldn't workers be expected to cheat and lay unfertilized eggs to produce sons of their own? It doesn't help that the sex ratios are skewed, because such a skew reduces the average expected reproductive output of females, something that also exactly cancelling the benefits from relatedness (West and Gardner 2010).

Another hypothesis is instead that strict monogamy, where females only mate with one male over her entire lifetime, is crucial for the evolution of eusociality. Such a system leads to the same coefficients of relatedness as in humans, where a worker is equally related to her offspring as her sisters. Any small efficiency benefit for rearing sisters instead of own offspring would over time favor eusociality. This hypothesis has empirical backing as the ancestral mating system for all species having evolved eusociality is monogamy. However, it fails to explain why workers in the eusocial hymenoptera are all females, or why workers feed female larvae with male eggs, facts that Hamilton's theory explain very elegantly.

Research is still ongoing on these topics. Note, however, that this mainly concerns the evolution of eusociality specifically, not Hamilton's theory of kin selection generally. This is crucial to remember when reading the next section.

A Challenge

Hamilton's work on kin selection has long been regarded as foundational for evolutionary biology. However, science wouldn't be science if the kin selection hegemony went unchallenged. If anything would bring lasting fame among evolutionary biologists it would be shooting down Hamilton's ideas on kin selection—these have iconic status and are therefore a very tempting target. Recently, an Austrian professor of evolutionary dynamics at Harvard University, Martin Nowak, tried to do just that: turn our understanding of kin selection on its head—and to top it off he did it through a close examination of eusociality.

Nowak and his co-workers—Nowak's graduate student Corina Tarnita and the famed ant researcher and sociobiologist EO Wilson [who wrote the once so controversial book *Sociobiology* (Wilson 1975)]—have put in the most whole-hearted attempt at a replacement theory. Not only do they maintain that eusociality has not evolved in the manner Hamilton described, but that kin selection in itself is a faulty approach, only valid under exceptional conditions.

In a very controversial article, Martin Nowak and his colleagues pointed out that you shouldn't really view non-reproductive worker ants as individuals, but as parts of the queen (Nowak et al. 2010). And why not, workers do not reproduce, after all, so shouldn't any understanding of ant sociality focus on the queens and not on the workers? A human body is separated into somatic (body) cells and reproductive cells—sperm and eggs—the ant queen and the workers could be viewed in the same way, with workers representing some sort of detachable somatic cells (but note that then it should be the males who become workers, who get all their genes from the queen—females get half their genes from their fathers and are thus not just somatic extensions of the queen).

In fact, Charles Darwin himself was thinking in similar terms when he wrote in the Origin of Species that 'Thus I believe it has been with social insects: a slight modification of structure, or instinct, correlated with the sterile condition of certain members of the community, has been advantageous to the community: consequently the fertile males and females of the same community flourished, and transmitted to their fertile offspring a tendency to produce sterile members having the same modification' (Darwin 1859).

Observations give some credence to this viewpoint that individual worker ants are about as unimportant to the colony as single cells in a human body. Exactly how unimportant can be seen in some Asian *Camponotus*-ants (related to horse ants) where individuals defend their nest using suicide bombings. The ants defending the nest run up to enemies and contract their muscles so violently that the walls separating body segments burst. This causes a sticky liquid to squirt out of the ant's body, catching their enemies in a sticky trap (Jones et al. 2004).

The 'explosion' neutralizes the enemy ants, but simultaneously kills the worker ant itself. The suicide bombing is beneficial for the rest of the individuals in the nest, and thus to the reproduction of the queen, and through this, beneficial for the genes instructing individuals to blow themselves up. The same mechanism explains bee stings, where the stinging bee pays with her life for the privilege of tormenting you. More extreme forms of self-sacrifice are hard to envision.

But according to Nowak and his co-workers, this is not about kin selection. Instead, they propose five key steps in the evolution of eusociality:

1. The formation of groups. This can occur for any reason, not necessarily kin-based groups.
2. Certain pre-adaptations. For example, solitary bees forced together solve problems cooperatively, because bees are pre-disposed to work on a task until it is finished and avoid tasks already completed. Since they solve the same task at the same place, they end up working on the same task together, despite having no coordination of work effort. Also, females in Hymenoptera feed or guard their larvae, so they stay by the nest.
3. The origin of genes for eusociality. Such a mutation may simply result in the non-dispersal of females so that they stay and reuse the nest instead of building a new one. The pre-adaptations in point 2 turn this non-dispersal into cooperation.
4. Emergent traits are naturally selected. Cooperation changes the conditions for the cooperatively breeding females, so new types of behaviors become favored by natural selection.
5. Multilevel selection (selection of genes, individuals and nests) shapes the life cycle and caste system. Because of the cooperative breeding between females, a new unit emerges—the nest—that becomes a target of natural selection that didn't exist before. Nest traits now become of central importance for the reproductive success of females.

All occurrences of eusociality in Hymenoptera have evolved in nesting species (Hunt 1999) and pre-adaptations have been documented, so the scenario seems reasonable enough (Nowak et al. 2010). Note that the relatedness of the breeding females in this model is a *consequence* of the non-dispersal, not a *pre-requisite* for the evolution of eusociality.

The mutation—transmitted via the queen (how else?)—that suppresses worker reproduction could be favored through selection on group level; colonies where workers don't reproduce would have an advantage over colonies where workers do reproduce (Nowak and Highfield 2011). You *have* to disregard the individual workers' interests in this step of the model, because an individual will *never* have an advantage (in biological terms) from not reproducing. So although the system starts with positing advantages for the stay-at-home daughters, in the end the workers become parasitized by their mother, the queen, for the good of the nest.

If you think this sounds like a form of group selection you are right, but the normal problems with group selection are, according to Nowak, solved via spatial proximity. Worker sterility, in this view, evolved because hymenopterans in groups where only one individual reproduced had an advantage over groups where all females reproduced. This theory can be rephrased in gene-centered language like this: queens with genes inhibiting the reproduction of 'their' somatic cells (workers) have an advantage over queens allowing somatic cell (worker) reproduction.

On closer inspection, there is nothing really new in Nowak and his co-workers' model of the evolution of eusociality. Hamilton's general theory of kin selection was used to explain how eusociality evolved, but only under certain ecological conditions; in lineages with extended brood care, with defense of a shared resource and a pre-existing nesting behavior, just as in Nowak's model. If the young are dependent on long parental care, and predation rates are high, then individuals who attempt breeding alone may not last long enough to see their offspring to adulthood (Queller and Strassmann 2009). The crucial empirical observation that Nowak and co-workers omitted, however, is that eusociality in all cases has evolved in monogamous lineages, clearly indicating the central role of relatedness in understand eusociality (Boomsma et al. 2011).

In fact, Hamilton pointed out that limited dispersal would be one potential path to cooperation, because it would tend to keep relatives together. In such cases, unconditional cooperation with whoever is next to you can be favored, but then because those individuals will likely be relatives (Hamilton 1964).

Kin selection can also explain things that Nowak and his co-workers' theory cannot, such as the competition between the queen and the workers about the sex ratio that was described above, where workers fed the larvae with male eggs to shift the sex ratio from ½ to their genetically preferred ¼. Hamilton's kin selection nicely explains this conflict between queen and workers just by invoking coefficients of relatedness.

But if the theory is old news, why is Nowak and his co-workers' theory so controversial? This comes down to their general dismissal of kin selection, where they claim that Hamilton's theory has explained almost nothing. Read Nowak's book *Supercooperators*, and it is as though kin selection almost doesn't exist in nature (Nowak and Highfield 2011).

In a remarkable reaction to this challenge from Nowak and co-workers, 137 evolutionary biologists co-authored an article specifying how the work of Nowak's group was incorrect in its dismissal and how much that has been explained by Hamilton's insights (Abbot et al. 2011). And there were many additional articles by other authors dwelling on specific errors in the Nowak et al. article.

Remarkably, the not too controversial eusociality model is presented at the forefront of Nowak and his co-workers' article, but interspersed with heavy-handed verbal dismissal of kin selection. The backing of this dismissal, however, is presented only on-line, in a highly technical and mathematically challenging 43-page appendix. Mathematically competent biologists have marveled at this appendix in that it 'misrepresents the components of Hamilton's rule' and 'does not provide any new theoretical tools or concepts to address the many exciting biological questions for future research on social evolution and structured populations' (Rousset and Lion 2011). These criticisms are mainly leveled at the general attack on kin selection from Nowak's group. Kin selection is still crucial for understanding animal cooperation.

Eusociality—Termites and Naked Mole Rats

There are other examples of totally self-sacrificial cooperation in the animal world—where individuals surrender their whole reproduction to others—eusociality—also outside Hymenoptera. Two examples are termites and naked mole rats. These species do not have the odd form of sex determination that Hymenopterans have, but have the same genetic system as humans. So how do we understand the evolution of eusociality in these groups?

One general observation has been made, that these animal groups go through periods of inbreeding and outbreeding, something which shifts the coefficients of relation from those of humans to greater relatedness among siblings (Bartz 1979, 1980). The way it works is that the female (the termite/rat queen) and the male (the termite/rat king—yes there are kings in these systems) are unrelated, but both are the product of inbreeding. This inbreeding comes about through limited dispersal—they mainly mate within the nest. In this type of situation, the kings and queens will be related to their offspring with $r = ½$, just as in humans.

However, the strong inbreeding means that once a king and queen will mate outside the nest, the queen will have the *same genes* on both pairs of all of her chromosomes, and the same applies to the king. All offspring of this pair will then turn out be genetically identical—they get *the same genes* from their mother and the *same genes* from their father. The coefficient of relation between full siblings will therefore in extreme cases be 1.

If you look at termites, they mate in several generations inside their nests and are therefore much inbred. When these inbred individuals then swarm, they meet other inbred individuals but with another inbreeding history. Note that any amount of temporary inbreeding is enough to increase the coefficient of relation from ½ between siblings, which favors the eusociality you find in termites (Trivers 1985). The same occurs in naked mole rats.

But science wouldn't be science if other explanations hadn't been proposed for the eusociality in termites (Thorne 1997). The problem is that eusociality has a single origin in termites. Such unique events are very difficult to analyze.

Of great importance to settle discussions about the evolution of eusociality in general is the observation that eusociality has *only and always* evolved in full-sibling families; families where the average relatedness of individuals to their siblings is *at least* as high as the average relatedness of individuals to their own offspring (Boomsma et al. 2011). Whatever else is necessary to cross the threshold to eusociality, kin selection is a crucial precondition.

And kin selection can explain many more occurrences of cooperation in nature, such as why Belding ground squirrels are more willing to risk calling an alarm in the presence of close relatives, why birds preferentially help at the nests of their parents, why cannibalistic tadpoles eat each other but spit out their relatives, and not the least why eusociality evolved in Hymenoptera, termites and naked mole rats. Family matters.

Kin Selection in Humans

From the explanation of termite and naked mole rat eusociality, speculating researchers have turned their attention to humans. One problem with human cooperation is that we cooperate more than can be explained just by ordinary kin selection—a problem we will investigate more closely in the second half of the book.

It has thus been suggested that humans during a period of our evolution were subjected to periodical inbreeding while roaming the savannah, just as in termites and naked mole rats. The reproduction may then have occurred mainly within groups and only seldom between individuals from different groups. This explanation has the potential to explain two peculiarities with humans simultaneously: our global genetic similarity and our extreme form of cooperation.

For such a numerous species, humans are remarkably genetically similar. According to some researchers this might be explained by a population bottleneck that humans went through between 100,000 and 50,000 years ago, although the evidence is currently too scarce to decide the question. According to the proposed hypothesis, the human population can at that time have been as low as 3000–10,000 individuals (Ambrose 1998; Rampino and Ambrose 2000). A possible cause of this bottleneck could be the gigantic eruption of the volcano Toba on the island of Sumatra 73,500 years ago, which temporarily caused severe climate changes (Rampino and Stephen 1993). However, further research has refuted this hypothesis in that it shows that the extent of climate change was much smaller than previously suggested (Lane et al. 2013). Molecular data indicate a more probable scenario was a long period—many thousands of years—of continuous low population size (Hawks et al. 2000).

Be that as it may, there exist, as we will see in later chapters, other potential explanations besides inbreeding of the odd human habit of excessive cooperation. The 'termite explanation' is therefore not really needed (Trivers 1985). Nevertheless, genetic investigations of contemporary hunters and gatherers show that individuals tend to be related within groups (cousins, on average), a pattern that does provide a small degree of support for the 'termite hypothesis' in humans (Bowles 2006).

Do we humans cooperate according to kin selection at all? The answer is an unequivocal *Yes*. Not all people on earth are privileged to have a family, but most are (and we all have one biological father and one biological mother). Despite the fact that family constellations can be very different (and still work well) most families are based on genetic kinship.

It is important to stress that genetic relatedness is not necessarily the best, or only, foundation for a family. On the contrary, as we shall see later, humans cooperate for many more reasons than genetic relatedness. Even in the nuclear family the two parents are the ones cooperating most with each other, despite them being the *least* genetically related. The explanation for this is, as we have seen, that parents have a common interest in having surviving offspring. But we cannot neglect the large importance of genetic relatedness for how our societies are structured.

A lot of people want to separate evolutionary explanations from humans. They seem to think that a biological explanation of unselfish behavior in some way diminishes the value of an action. But how true is such an assertion?

Humans have strong empathic feelings when we see someone suffering. We can put ourselves in the suffering person's situation and feel their pain. We can let this empathy motivate our actions and sacrifice part of our wellbeing for others. And these, not genes, are the type of feelings that drive human cooperation, mainly within the family but also with others, right?

All this is of course true. But the fact that feelings are such strong motivating factors is the exact reason that emotions are a prime target for natural selection. How would our genes make their 'wishes' known to our brains other than through strongly motivating feelings? These types of social emotions—love, empathy, belonging, lust, parental love—are not the answer to why we cooperate. On the contrary, they are part of the question, part of what needs to be explained.

Emotions are the proximate, closest explanation of cooperation. What we are looking for is the ultimate cause—the cause that explains *both* our social emotions *and* our cooperation.

But many people shy from such explanations. Especially during the 1970s, when researchers were beginning to understand kin selection and other genetic causes behind human altruism, the resistance was almost scary. Proponents of biological explanations were called resisters and Nazis, determinists and social Darwinists—despite the fact that most researchers didn't hold these views (Segerstråle 2001). This resistance is not unlike the type of resistance by the church against Darwinism.

> The Church swarms with people who have no spiritual sinew, and whose lungs cannot breathe the invigorating air of Truth: they take up the cry of that timid and decorous spinster who, on hearing an exposition of the Darwinian theory that men are descended from apes, said, 'Let us hope it is not true, or if it is, let us hush it up.' (Unclear origin—Quote investigator)

But why, really? The observation that humans are altruistic and cooperative because of our genes, that our actions have natural explanations, in no way diminishes the values of our actions. Unselfish behavior is worth just as much whether it is caused by nature or nurture. Why not turn things around and be happy that unselfishness is part of our very nature?

Surely, cooperation within families is a strong norm. The overwhelming majority of societies on earth have the family as the basic unit. It is even written into the UN Universal Declaration of Human Rights, Article 16 (3).

> The family is the natural and fundamental group unit of society and is entitled to protection by society and the State.

Most families are constructed around genetic relatedness. Parents, who are not genetically related to each other but to their common children, cooperate for egoistic reasons in that they both benefit. But there are many different types of families. If it isn't mother-father-children, which it most certainly isn't in all parts of the world, it can be brother-sister-sister's children, or mother-mother-mother-father-children, or mother-children, or father-father-children. If you lift your gaze above what in the western tradition is termed the nuclear family, you will find grandmothers, grandfathers, cousins, second cousins, more distant relatives and clans.

From a biological perspective this type of cooperation makes perfect sense. The cause for this level of cooperation is to be found in the evolutionary process that Bill Hamilton discovered when describing kin selection. Genetic relatedness is a crucial factor for understanding human cooperation.

But is it the *whole* explanation of *all* cooperation? Most certainly not. What you do when you go to a café to buy a cup of coffee and a doughnut is cooperation that is so far beyond the type of cooperation explained by kin selection that it demands a completely different explanation. In the next chapter we will meet another genetic explanation that helps to explain other forms of cooperation. And in later chapters we will find even more explanations.

Note that *all* other cooperation between animals besides kin selection needs to benefit the acting individual directly, or the genes that it is carrying; it is only kin selection that works through genes coding for self-sacrificial behavior helping copies *of themselves* in other individuals. *All* other genetically coded self-sacrificial behavior has to benefit the acting individual itself or its own genes. (Though when we introduce culture towards the end of this book this will allow for more options.)

Now—friendship: are there evolutionary conditions allowing for friendship?

References

Abbot, P., et al. (2011). Inclusive fitness theory and eusociality. *Nature, 471*, E1–E4.

Ambrose, S. H. (1998). Late Pleistocene human population bottlenecks, volcanic winter, and differentiation of modern humans. *Journal of Human Evolution, 34*, 623–651.

Bartz, S. H. (1979). Evolution of eusociality in termites. *Proceedings of the National Academy of Sciences, 76*, 5764–5768.

Bartz, S. H. (1980). Correction. *Proceedings of the National Academy of Sciences, 77*, 3070.

Boomsma, J., Beekman, M., Cornwallis, C. K., Griffin, A. S., Holman, L., Hughes, W. H. O., et al. (2011). Only full-sibling families evolved eusociality. *Nature, 471*, E4–E5.

Bowles, S. (2006). Group competition, reproductive leveling, and the evolution of human altruism. *Science, 314*, 1569–1572.

Colman, A. M., Browning, L., & Pulford, B. D. (2012). Spontaneous similarity discrimination in the evolution of cooperation. *Journal of Theoretical Biology, 299*, 162–171.

Darwin, C. (1859). *On the origin of species by means of natural selection, or the preservation of favoured races in the struggle for life* (1st ed.). London: John Murray.

Dawkins, R. (1976). *The selfish gene*. Oxford: Oxford University Press.

Dugatkin, L. A. (1997). *Cooperation among animals: An evolutionary perspective*. Oxford: Oxford Series in Ecology and Evolution.

Gardner, A., & West, S. A. (2010). Greenbeards. *Evolution*, 64, 25–38.

Grafen, A. (1985). A geometric view of relatedness. *Oxford Surveys in Evolutionary Biology, 2*, 28–89.

Hamilton, W. D. (1964). The genetical evolution of social behaviour I and II. *Journal of Theoretical Biology* 7: 1–16 & 17–52.

Hamilton, W. D. (1970). Selfish and spiteful behaviour in an evolutionary model. *Nature, 228*, 1218–1220.

Hawks, J., Hunley, K., Lee, S. H., & Wolpoff, M. (2000). Population bottlenecks and Pleistocene human evolution. *Molecular Biology and Evolution, 17*, 2–22.

Hughes, W. O. H., Oldroyd, B. P., Beekman, M., & Ratnieks, F. L. W. (2008). Ancestral monogamy shows kin selection Is key to the evolution of eusociality. *Science, 320*, 1213–1216.

Hunt, J. H. (1999). Trait mapping and salience in the evolution of eusocial vespid wasps. *Evolution, 53*, 225–237.

Jones, T. H., Clark, D. A., Edwards, A. A., Davidson, D. W., Spande, T. F., & Snelling, R. R. (2004). The chemistry of exploding ants, *Camponotus* SPP. (*Cylindricus* COMPLEX). *Journal of Chemical Ecology, 30*, 14791–14492.

Jorde, L. B., & Wooding, S. P. (2004). Genetic variation, classification and 'race'. *Nature Genetics, 36*, S28–S33.

Lane, C. S., Chorn, B. T., & Johnson, T. C. (2013). Ash from the Toba supereruption in Lake Malawi shows no volcanic winter in East Africa at 75 ka. *Proceedings of the National Academy of Sciences, 110*, 8025–8029.

Nowak, M. A., & Highfield, R. (2011). *Supercooperators*. London: Canon Gate.

Nowak, M. A., Tarnita, C. E., & Wilson, E. O. (2010). The evolution of eusociality. *Nature, 466*, 1057–1062.

Passera, L., Aron, S., Vargo, E. L., & Keller, L. (2001). Queen control of sex ratio in fire ants. *Science, 293*, 1308–1310.

Queller, D. S., & Strassmann, J. E. (2009). Beyond society: The evolution of organismality. *Philosophical Transactions of the Royal Society B: Biological Sciences, 364*, 3143–3155.

Rampino, M. R., & Ambzrose, S. H., (2000). Volcanic winter in the Garden of Eden: The Toba super-eruption and the Late Pleistocene human population crash. In F.W. McCoy & G. Heiken, (Eds.), *Volcanic hazards and disasters in human antiquity. Geological Society of America Special Paper* (Vol. 345, pp. 71–82).

Rampino, M. R., & Stephen, S. (1993). Bottleneck in human evolution and the Toba eruption. *Science, 262*, 1955.
Rousset, F., & Lion, S. (2011). Much ado about nothing: Nowak et al'.s charge against inclusive fitness theory. *Journal of Evolutionary Biology, 24*, 1386–1392.
Segerstråle, U. (2001). *Defenders of the truth: The sociobiology Debate*. Oxford: Oxford University Press.
Sherman, P. W. (1981). Kinship, demography, and belding's ground squirrel nepotism. *Behavioral Ecology and Sociobiology, 8*, 251–259.
Thorne, B. L. (1997). Evolution of eusociality in termites. *Annual Review of Ecology and Systematics, 28*, 27–54.
Trivers, R. L. (1985). *Social evolution*. Menlo Park, CA: Benjamin/Cummings.
West, S. A., & Gardner, A. (2010). Altruism, spite, and greenbeards. *Science, 327*, 1341–1344.
West, S. A., El Mouden, C., & Gardner, A. (2011). Sixteen common misconceptions about the evolution of cooperation in humans. *Evolution and Human Behavior, 32*, 231–262.
Williams, G. C. (1966). *Adaptation and natural selection*. Princeton: Princeton University Press.
Wilson, E. O. (1975). *Sociobiology: The new synthesis*. Cambridge: Harvard University Press.

Friends

Abstract Reciprocal altruism and the prisoners' dilemma are explained and it is related that there are very few examples of these processes in nature. They do seem to exist in some species of primates, however, and have given rise to "the social brain hypothesis" indicating that sociality has driven the evolution of the primate brain. Finally, conditions for cooperation to arise through indirect reciprocity, network reciprocity and group selection are outlined.

Bill Hamilton showed how self-sacrificial behavior between related individuals can emerge as the result of a simple biological mechanism—genes 'help themselves' in other individuals. Or, to put it differently and more correctly, genes that hold instructions for self-sacrificial, altruistic behaviors targeted at related individuals carrying the same genes are favored by natural selection. The key to the success of the explanation came through Hamilton's change of perspective: instead of analyzing individuals he looked at how things would work on the genetic level.

The person who formalized the next explanation of self-sacrificial cooperation was Bob Trivers, who was inspired by Hamilton's progress but understood that more needed to be done. Trivers would also go on to become one of the giants in the field of evolutionary biology. His five first articles came at a rate of one per year and all provided central insights within five different areas: reciprocity (Trivers 1971), parental investment (Trivers 1972), sex ratios (Trivers and Willard 1973), parent-offspring conflict (Trivers 1974), and social insects (Trivers and Hare 1976).

This remarkably productive phase of Trivers' career ended when a psychological breakdown hospitalized him. Trivers was diagnosed with bipolar syndrome and has since suffered a series of breakdowns (Brown 2005). He jokes about his current field of enquiry—self-deception (Trivers 1991)—by retelling his own experiences: 'I had to be tied down on a back ward of a mental hospital for six months for it to occur to me that I am doing less well than the average of my colleagues.' (Virtanen 2005) But between these breakdowns Trivers is a brilliant researcher.

The problem that Trivers wanted to solve in the 1970s was a glaring inconsistency in current biological theory. Each and every one of us has daily interactions with people that we are not related to. Yet Bill Hamilton's explanation of

P. Lindenfors, *For Whose Benefit?*, DOI 10.1007/978-3-319-50874-0_6

cooperation solved cooperation only between relatives. Trivers went on to investigate if there was some way to biologically explain also self-sacrificing cooperation between individuals who were *not* related.

Trivers realized that what we are doing in such interactions is trading favors so that one individual makes a sacrifice first, to which the other individual then reciprocates. Mutual favors; I scratch your back, you scratch mine. Just as with many other brilliant insights, this suggestion seems almost painfully self-evident in hindsight. The idea dates back to Hume and had been analyzed in economy before Trivers put the idea on solid theoretical footing also in biology (West et al. 2011).

The problem to be solved was how such an exchange of favors can get started. One of the participants must start the interaction and thus trust the target individual enough to provide the first favor. The receiving individual must then reciprocate. But how much—and when?

The Prisoners' Dilemma

In the situation described, a dilemma involving two individuals emerges. The first individual (A) has a chance to help the second (B). Since A has the possibility to help, one can safely assume that A has something that B doesn't; food, for example. Because A has a surplus and B has a shortage, the food will be more valuable for B, who perhaps is starving, than to A, who has more than enough. If the situation is the same, but reversed, in a future interaction, both individuals will have provided cheap aid but received valuable help.

The problem to solve is that A initially will gain in *refraining* from helping B, since A then gets to keep the food. However little it is worth at that very moment, it still is worth something. And in the next stage, B can simply refrain from reciprocating. Then B will benefit from the interaction since B is helped *and* gets to keep the food. So how can reciprocity work?

There is also another complication, termed the 'banker's paradox': if B is in *too* bad of a situation, A cannot really expect to be paid back later—B can just die. So just when you are in the greatest need of help, a system of exchange could be expected to collapse (Tooby and Cosmides 1995; Krupp et al. 2011).

Conditions such as these set up something termed *the prisoners' dilemma*—a famous problem in a branch of mathematics termed Game Theory. The mathematical version of the dilemma was first created in the 1950s by Merrill Flood and Melvin Dresher at the Rand Corporation think tank, founded in the US after the Second World War to try to predict and understand the behavior of the Soviet Union. It was, however, the mathematician Albert W. Tucker who gave the game its name when trying to explain the game in a simplified way and framed it as a dilemma between two prisoners (Kuhn 2014).

Two prisoners, A and B, are given the same choice. Each can either testify against the other or cooperate by keeping quiet. Both prisoners are informed that the other has the same choice. If A testifies while B stays quiet, A goes free while B

Table 1 The prisoners' dilemma

	B keeps quiet	B testifies
A keeps quiet	Both get 1 year in prison (R = 1, R = 1)	A gets 10 years in prison, B goes free (S = 10, T = 0)
A testifies	A goes free, B gets 10 years in prison (T = 0, S = 10)	Both get 4 years in prison (P = 4, P = 4)

The letters in parentheses indicate the payoffs for each player in each interaction. The first letter in each parenthesis indicates the payoff for player A while the second letter indicates the payoff for player B. The letters mean R = 'Reward': the reward for mutual cooperation; S = 'Sucker's payoff': the punishment for having been cheated; T = 'Temptation': the temptation to cheat; P = 'Punishment': the punishment for mutual defection. The formal conditions are that $T > R > P > S$

gets a long prison sentence, and vice versa. If both testify, they both get a sentence of medium length. If they both keep quiet, both get short sentences. These are the payoffs in table form (Table 1).

If both prisoners cared about each other they would of course both remain silent and each get one year in prison, which sums up to two years in total. This is the lowest total sum punishment possible in the game, and equally distributed over both players, thus also fair. The problem, of course, is the temptation to cheat. If you are the only one cheating you go free—wouldn't you want that?

How should this problem be approached? Let's look at it emotionally detached, from a purely rational standpoint. Imagine that you are player A. If B keeps quiet you gain from defecting since you will go free instead of spending a year in prison. On the other hand, if B chooses to defect, then you *also* gain from defecting since you will end up spending less time in prison—four years instead of ten.

This implies that it is individually beneficial to always defect! No matter what B does, you always gain from defecting. That wasn't a very hard dilemma to solve, was it? If we go back to the problem of the evolution of cooperation this means that it would be *impossible* for self-sacrificing cooperation to evolve between non-relatives. You *always* gain from *not* helping.

What is amiss? After all, humans cooperate on a daily basis with people who are not relatives. This is a fact any one of us can recognize just by going out our front door and looking at our fellow citizens. How has this cooperation evolved?

The crucial factor is time. Cooperative interactions must be *repeated*—in the example above we just looked at a single interaction. If you play the prisoners' dilemma with the same person repeatedly it will be advantageous to cooperate instead of defecting, will it not? The total time in prison will then be the lowest possible for both individuals. You will keep ending up in prison for one year (both cooperate) instead of four (both defect).

But is there no way to maneuver yourself into some small advantage? If, for example, you manage to defect just sometimes, you would gain an even larger benefit. Maybe there is a cunning formula to follow to minimize your own years in prison?

But the same applies to the other player, also attempting to minimize years in prison. How do you guard against that?

Trivers pointed out that in addition to repeated interactions it is a crucial prerequisite that you *recognize and remember* the other player. If you remember who defects and don't cooperate with them—there is after all a kind of 'market' of potential players to choose from (Bshary and Noë 2003; Noë and Voelkl 2013)—then this type of self-sacrificial cooperation will spread in a population through natural selection. Someone who defects will not be part of any future cooperation.

But what if you can cheat the other player to believe that you always cooperate but sneak in an extra reward just every now and then, for example as the very last thing you do? Is there no way to cheat Trivers' system?

Two researchers, the political scientist Robert Axelrod and the evolutionary biologist Bill Hamilton (whom we already know) took this one step further. In 1981 they set up a computer tournament where they invited anyone interested to submit a strategy for repeatedly playing the prisoners' dilemma. Axelrod and Hamilton got 15 strategies that they proceeded to turn on each other in an imitation of the evolutionary process where different genetic strategies would compete against each other. The strategies that did best were selected to compete against other strategies.

Many game theorists submitted strategies, each more advanced than the previous. There were strategists that modeled the other player's behavior and used mathematical decision formulas to calculate the next step. Further, strategies were entered in the competition that cooperated and defected randomly, and there were strategies that always defected, or never defected.

The winning strategy was one of the simplest of them all: *tit-for-tat*. This strategy had only two rules: first, cooperate. Then at every turn, do what the other player did in the previous round. In this way *tit-for-tat* cannot be taken advantage of.

To intensify the competition Axelrod and Hamilton published the winning strategy and asked for new contributions. This time 62 new strategies were entered, from six countries. Most entries came from computer programmers, but also from biologists, economists, and physicists. Again *tit-for-tat* competed; again it won.

Axelrod and Hamilton reported that *tit-for-tat* beat all other strategies and summarized the strategy's advantage in three points (Axelrod and Hamilton 1981):

1. Never be the first to defect.
2. Only defect when the other player has defected—this is punishment.
3. Be forgiving after only having punished your opponent once. This is an optimistic trait: Always trust your opponent after a single invitation to cooperate.

Researchers have later noted that the history of this tournament isn't as clear as in this description and that there were other stable strategies (Binmore 2007). *Tit-for-tat* is, for example, not forgiving of mistakes, but can end up in an infinite vicious circle of retaliation after another player has defected only once. More advanced strategies have therefore been constructed later, such as *tit-for-two-tats* (that punishes only after the opponent has defected twice), *generous tit-for-tat* (that forgives defection with a certain probability), and *win-stay, lose-shift* (that repeats its own previous action if that interaction was successful, otherwise switches).

The big picture is still that which Axelrod and Hamilton reported, however: the most stable way to get working cooperation in an interaction between two parties where there is an initial cost is to decide what to do based on the above three points.

Actually, at heart reciprocity may depend on the same cost/benefit-analysis as in Hamilton's rule—Hamilton's formula has been reformulated to cover a broader selection of cooperation by David Queller. What Queller did was to replace relatedness in Hamilton's equation with a measure of covariance in the population between the phenotype of individuals performing the altruistic act and the genotype of the recipients. If the benefits of the altruistic acts with a greater probability falls on those who are likely to pass the altruistic behavior on, then altruism survives (Queller 1985).

This formulation shows that self-sacrificial cooperation will spread if the benefits from the altruistic acts favor individuals that will pass the behavior on, no matter relatedness. This covers relatedness as well as reciprocity, even between different species (Fletcher and Zwick 2006).

Examples from the Animal World?

Trivers had formulated a new simple rule explaining costly cooperation, this time between individuals who were unrelated: *reciprocal altruism*. But does this type of cooperation exist in nature? Researchers immediately set out to investigate this question. One example of reciprocal cooperation between unrelated individuals, that became a classic, was reported in vampires—but vampires from Central and South America, not from Transylvania: vampire bats.

A biologist by the name of Gerard Wilkinson noted that vampire bats fulfill all criteria for reciprocity. They live in stable groups where all individuals know each other and thus have the possibility of re-paying any help. And vampire bats need help rather often (Wilkinson 1990).

Vampire bats can survive only three days without food. A third of young bats (under two years) and 7% of older bats fail to find food on any given night. Assuming that all bats fail equally often, this means that normally, 10% of the bats below two years of age and 0.5% of the older bats fail to get food on any random two nights in a row and therefore are in desperate need of food, or have offspring in desperate need of food—vampire bats feed each other's young. Without any system for giving and getting help, these bleak statistics mean that there would be no vampire bats if mutual aid didn't exist.

A bat that has just eaten will have more food than it really needs. So it will be relatively cheap for it to regurgitate some blood for other group members or off-spring, and very valuable for the individuals who haven't eaten to be helped. Wilkinson showed that the bats exchange feedings with each other on a regular basis. Those having eaten regurgitate blood to aid those who haven't eaten. Those who are helped pay back another time.

Another example of *tit-for-tat* was presented by the German biologist Manfred Milinski through a series of incredibly clever experiments on stickleback fish. These fish have the peculiarity that they inspect predator fish that come too close to their nests. Given the opportunity, they do this together with neighboring sticklebacks. The problem that emerges is that no stickleback wants to be the one coming closest to the predator since it then runs the greatest risk of being eaten.

What Milinski did was to put models of predator fish in front of a stickleback in an aquarium. He then put a mirror on the side, something which deceived the stickleback to believing that another individual was also examining the predator. If Milinski angled the mirror so that it appeared that both 'companions' went equally close to the predator, then the stickleback would continue its inspection. But if Milinski angled the mirror so that it seemed as if its 'companion' stayed a little behind, the stickleback ended the inspection and retreated to its nest. This type of punishment is just what you would expect according to the *tit-for-tat* model of cooperation (Milinski 1987).

For the evolution of mutual cooperation, cooperating individuals don't even have to be the same species. Another example of mutual cooperation comes from cleaner wrasse fish and their 'customers'. The customer fish that is to be cleaned will come to the place where the cleaner wrasse are and stay still, often with its mouth open. The cleaner wrasse will proceed to pick away dead tissue and parasites.

The customer at this time has the opportunity to eat the cleaner wrasse. The cleaner wrasse, on the other hand, has the opportunity to take bites out of the customer. But neither of them do this and cooperation is maintained through an exchange of favors.

These types of examples, and there were many, were for a long time regarded as clear examples of reciprocity in the animal world. But science would not be science if these examples weren't challenged.

The problem with the vampire bats is that it wasn't possible to confirm that the interactions between the bats were repeated pairwise so that the provider also got the favor returned later by the same individual. Additionally, the individuals in the vampire bat groups were all related—cousins, on average. Thus, Hamilton's theory of kin selection was sufficient to explain the bats' food sharing.

There is also another potential explanation for the regurgitation of blood. Hungry vampire bats beg so intensely that it may be cheaper for the sated individual to pay some blood just to get the pestering to end. Maybe this helping system therefore is a better example of reciprocal parasitism where hungry individuals take turns to extort food from each other?

Concerning the sticklebacks, the helper behavior provides no special advantage to the fish in reproductive terms. And in the cleaner wrasse, the interaction is actually not an example of reciprocity, but of mutualism—both parties benefit from the interaction immediately, as in lichen, if not as developed. If there is no cost involved, there is no evolutionary riddle to be solved—we expect organisms to do what benefits them directly!

So this turned out more difficult than it seemed at first sight. What does it really take to scientifically show the existence of reciprocity in nature? In 2009 a zoology professor at Cambridge, Tim Clutton-Brock, put together a list of four conditions that need to be met.

1. Reciprocity—the same individuals have to help each other time and again.
2. Fairness—the self-sacrificial behavior has to be about equally distributed between the two reciprocating individuals.
3. Not kin selection—the cooperating individuals cannot be related or potential partners.
4. Evolutionary costs and benefits—the self-sacrificial behavior has to be costly *in reproductive terms* for the provider and beneficial *in reproductive terms* in the recipient.

Clutton-Brock reviewed all reported examples of reciprocity and not found a single example of cooperation in nature that fulfills all four of these criteria. The best examples he found were when individuals in primate societies or deer herds help each other removing parasites. Primates and deer trade these favors, so that the individual removing parasites expects reciprocation at a later stage. In this way, trust is established between pairs of individuals trading such parasite cleansing (Clutton-Brock 2009).

According to Clutton-Brock, what is missing from these last examples is the cost. No major effort is involved in removing parasites from other individuals. Above all, there is no cost involved *in reproductive terms*—and that is what counts in evolutionary biology. It appears that the cost of carrying ticks and lice isn't that high.

Without any reproductive costs or benefits this kind of cooperation isn't subject to natural selection and without natural selection there is no adaptive evolution. It is survival and reproduction that count. This is why the example of the vampire bats, had it checked out, would have been so great. Clutton-Brock's review does not exactly prove that reciprocity doesn't exist in animals—he is simply pointing out that no one yet has pointed to a single example that withstands rigorous scientific scrutiny.

As the American primatologist Joan Silk has reminded us, however, absence of evidence is not evidence of absence. It may well be that we just haven't found the relevant examples. Nevertheless, it seems clear that reciprocity is much less common in other organisms than it is in humans (Silk 2013).

However, the cases of reciprocity that come closest to fitting Clutton-Brock's criteria are interesting in their own right. The central point here is that you can't simply look at behaviors outside their long-term context. Primates removing parasites from each other are not just trading favors, but trading favors in ways that result in lasting social bonds. Or, to put a human label on the relationship that ensues: they become friends. And friendship seems very important in some primate societies (Van Schaik and Kappeler 2006).

Friends can provide support in conflicts, share food—and remove parasites. Over the long term, such friendships result in measurable effects on individual health. It has for example been reported that males in baboon bands who have stable cleaning relationships with females live longer and produce a higher number of surviving offspring than baboon females that do not partake in such friendships (Silk et al. 2010). What remains to be shown for this example to hold up to scientific scrutiny is causality—it can just as well be the case that baboons prefer to strike up friendships with individuals who are already healthy.

It seems as if individuals who are not part of lasting exchanges of favors pay a reproductive cost. It is just that the time frame is long—longer than the average field study. A single interaction may mean very little and cost nothing, but the sum of interactions over a lifetime can mean the difference between life and death.

One of the noticeably different traits of primates—the order of mammals of which humans are a part—is our high degree of sociality. This is not to say that all monkeys and apes live in social groups, or that no other animals live in complex social groups—there are very intricate social systems also in carnivores (for example lions, wolves and African wild dogs)—and there are also species of primates where individuals live in small family units or solitarily, for example orangutans.

The special thing with primates is instead the diversity of social systems. For example, baboons live in groups where there are always fewer males than females. Adult baboon males get in each other's faces and therefore prefer the company of submissive elderly or immature males. Other males wait on the periphery for their chance to take over.

While adult males guard their group against other males—baboon males have huge, knife-like canines that they use on each other—the females organize their own hierarchy. Newborns get their mother's rank. Social mobility is by family, not individually.

Other social systems, each with specific problems, exist in langurs, macaques, howler monkeys, colobus monkeys, leaf monkeys, chimpanzees and gorillas. These species all live in social groups with complex social structures, both unique and similar to those of other species.

Thus, many primates have well developed social skills. Like many other species they can recognize each other and distinguish kin from non-kin. But they also seem to be able to assess the value of interactions (resources and favors), remember interactions with other group members and distinguish between cooperating and non-cooperating individuals. They also have a sense of the social relationships between other individuals.

In an experiment with vervet monkeys, researchers recorded screams from different infants and played them back to the group. The adults looked not only in the direction of the screaming, but also at the mother (Cheney and Seyfarth 1980). They registered the infant's call but also comprehended the social ties that were supposed to be the solution to the problem.

Aside from baboons, chimpanzees are the species where scientists have discovered most about the effects of long term cooperation. In chimpanzees it is the males who stay while females move between groups. It is therefore between males you expect most cooperation in chimpanzees, for reasons of kin selection if nothing else.

But chimpanzee males also cooperate with unrelated males by trading favors like parasite cleaning, support in conflicts and sharing prey. They can also trade *different* favors, such as cleaning for conflict support, or cleaning for meat.

It has further been shown that chimpanzees have many of the cognitive traits needed to remember reciprocal exchanges of this kind. They can, to a degree, suppress immediate needs, judge quantities and distinguish between purposeful and non-purposeful behavior (Mellis and Semmanm 2010).

Some researchers even claim that monkeys may have a sense of justice. Two biologists, the American Sarah Brosnan and the Dutchman Frans de Waal, discovered that if you give two capuchin monkeys (de Waal 1989, 1997) or chimpanzees (Brosnan et al. 2005) a task to solve together and then give one of them a nicer reward (grapes) than the other (cucumber), the monkey who gets the bum deal gets upset and refuses to collaborate again.

However, later research has shown that they mostly seem to get upset over not getting the nicer reward—not that it's the other individual's nicer reward. They become equally upset if you make them carry out a task and as a reward give cucumber while showing only the grapes, with no other individual near (Tomasello 2009). So whether primates (except humans) have a sense of justice is still unclear.

The Social Brain

Considering all the interactions that the members of a primate social group have to keep in mind, both their own and those of others, it isn't strange that the number of possible social interactions in primate groups has turned out to be intimately associated with the organization of the brain. In species that typically have a larger number of individuals in their social groups—and where there is therefore a greater number of potential relationships—the size of the neocortex, the executive part of the brain that is used to handle 'higher' cognitive functions, is larger than in other species.

To cooperate with relatives does not demand an especially advanced brain, but if you have to juggle kinship, reciprocal friendship and other relationships with many individuals over large time spans, it becomes complicated on a whole other level. Every time a social relationship changes, this changes the whole group's relationships. In big complex groups, the brain's capacity is therefore decisive for how well you can navigate the social network.

An important explanation of the brain's evolution in primates is therefore the high degree of sociality in this order. To get food is one thing, but *all* species need to be good at finding resources. In complex social groups, on the other hand, the brain is much more important.

A big brain is something that many humans hold as a sign of progress. This according to the assumption that evolution will always favor the larger brain since smartness increases the chance of survival as well as the ability to attract a partner. If you view evolution in this way, humans are the self-evident pinnacle of evolution—first among the first! (Primate means 'first' in Latin).

But nothing could be farther from the truth. The brain is an expensive organ to maintain and doesn't always provide an advantage. In humans, about 20–25% of the daily energy budget goes to running the brain, while in other primates the proportion is about 5%. There must be a good reason to evolve such an expensive organ.

Another of the pinnacles of evolution, a group of animals that has evolved for exactly the same amount of time as humans but that has found a completely different solution to life's challenges are the tunicates—animals that spend their whole adult life stationary, attached to the substrate, filtering water for food. At mating time, the males release their sperm straight into the ocean and with a little luck some sperm find their way to a tunicate female and fertilize her eggs.

Once the eggs hatch, small tadpole-like larvae swim away into the ocean to find a suitable place to spend the rest of their lives. These larvae are equipped with a brain that governs both movement and sense of direction—a primitive brain, but a brain nonetheless. When a tunicate larva finds a suitable place to live, it attaches to the substrate, and transformation to adulthood commences.

A stationary filtrating tunicate has no need to govern movement or of any senses. So the tunicate promptly digests its own brain and thus recycles the energy invested. As adult, the tunicate lives a completely stationary life without a brain. (Teenagers may recognize their parents here.) Despite having the genes to make a brain—and despite having a brain as a juvenile—none of this is thus utilized in the adult form. Extreme? So are humans.

To see humans as the pinnacle of evolution is just as correct as seeing tunicates as the pinnacle of evolution, no more, no less. Both of us have evolved for the same amount of time from our last ancestor common to both of us, and both of us have adapted remarkably well to our environments. We humans have merely arrived at a somewhat different solution—but in an evolutionary perspective this is true for *all* extant organisms; we are *all* the pinnacles of evolution.

So we have to be careful in assuming that a big brain is a measure of success. It happens to be true in humans, to be the best solution to *our* challenges, in *our* sub-section of the diversity of life—primates, and specifically humans. Just as well, otherwise you wouldn't be able to read this.

Nevertheless, the strong association between the size of the neocortex and social complexity in primates has led the British anthropologist Robin Dunbar to turn the question around and try to calculate just how large a social group a human brain can handle. If our brains are just overgrown primate brains, the maximum size of a human social group that a human brain can handle should be about 150, a number now termed as Dunbar's number (Dunbar 1992).

But our brains are not 'just' scaled-up versions of ordinary primate brains, we are capable of completely different feats than other primates because we use our brains for culture (more on that later). Humans have brains that are so ridiculously large already at birth that human births are described as a divine punishment in the Bible. Our brains are capable of such feats as learning languages, composing atonal music, conversing about the weather and solving differential equations. We can do stuff with our brains that other primates cannot even dream of dreaming of.

The general reasoning behind Dunbar's number depends crucially on the assumption that primate social units are uniquely limited by species' cognitive abilities, something that just isn't the case—there are a number of other factors that limit social unit size, such as food availability and predation, meaning that a relationship between cognitive capacity and social group size systematically underestimates maximum social group size.

Further, the differences between group sizes of chimpanzees and human hunter-gatherers are not as predicted by the theory. Empirical studies of human networks have indicated that human social groups in general come in a large variety of sizes, many much larger than 'Dunbar's number'.

Despite its many limitations Dunbar's number has become very popular, mainly outside science. For example, in 2007 the Swedish Tax Department re-organized their work force so that no work group would be larger than 150 people (TT/The Local 2007). The employees in the department indignantly (and rightly) protested being treated as nothing more than overgrown apes. Studies of real human networks give a more reasonable empirically derived number at about 250–300 (McCarty et al. 2001).

The human brain is oddly capable of incorporating an operating system (language) and programs (socially learnt knowledge) and is thus an exception (we'll get back to that). Nevertheless, there is no doubt that there is correlation between the size of the social group and the size of the neocortex among primates. Interestingly enough, there are also signs of a difference between the sexes in this regard. My own contribution to this line of brain research has been to show that if you look at males and females separately, the correlation between social group size and neocortex size applies only to females. In males, the size of groups has no correlation whatsoever with neocortex size (Lindenfors et al. 2007).

What is *that* about? It turns out that what matters most for male reproduction is not only how to navigate intricate social networks. In species where males compete intensely over females other parts of the brain develop more; those that have to do with physical competition.

Males and females have specialized roles in primate groups. Females have close cooperative interactions with each other in order to maximize their number of surviving offspring. Males are instead drawn together around females to compete physically for their favors and to keep other males away. These differences have left different traces in the brains of both sexes (Lindenfors et al. 2007).

How does this translate to humans? It would lead to the testable prediction that females are more socially adept while males are more physically competitive, on average (and only on average—individual variation is, as always, expected). And at least when it comes to male violence it would seem to bear out from the observations that 90–95% of the world's prisoners and nearly all soldiers are male (Lindenfors and Tullberg 2011). But one must be very careful to point out that this is descriptive, not a prescriptive pattern—nature doesn't tell you what is right and wrong behavior.

Other Possible Genetic Explanations of Cooperation

But do you really have to reciprocate *yourself* all the time in the manner Trivers suggested? Isn't it foolish to learn only from *your own* interactions with other individuals? Now that we possess this impressive brain due to all the interactions going on around us, wouldn't we expect that individuals are able to learn also from *other individuals interacting with each other*?

The biologist and mathematician Martin Nowak—the same researcher who is currently trying to topple Hamilton's theory of kin selection—has taken Hamilton's and Triver's research further and mathematically formalized what would be necessary for other forms of cooperation to evolve, in other social contexts than within families and in reciprocal interactions between pairs of individuals.

With five formulas, all expressed in the same way as Hamilton's rule (r > C/B), Nowak checks off all possibilities of genetic evolution of self-sacrificial cooperation that are feasible. In all formulas, *C* is the cost and *B* the benefit of the interaction, but the coefficient of relatedness is replaced by different criteria (Nowak 2006). The formulas are shown below.

- $r > C/B$: *Hamilton's rule*, where r is the coefficient of relatedness.
- $w > C/B$: *Direct reciprocity*, where w is the probability that two individuals interact again.
- $q > C/B$: *Indirect reciprocity*, where q is the probability of knowing someone's reputation.
- $k < B/C$: *Network reciprocity*, where k is the average number of neighbors in the network.
- $1 + (n/m) > B/C$: *Group selection*, where n is the maximum group size and m is the number of groups.

The two topmost explanations of cooperation should be familiar by now: kin selection and reciprocity. The bottommost formula describes the conditions under which group selection can evolve. It takes some rather special conditions, but if these are in place it is actually possible. But are these conditions too odd for group selection to be feasible in reality? The answer seems to be a cautious 'possibly'—group selection may actually exist in nature, but *if* it does it is rare—so rare it has yet to be observed (depending on what one means by 'group selection', of course).

Researchers have only just started to investigate network reciprocity in human societies, so we can expect progress—we will briefly discuss networks in the second to last chapter.

But indirect reciprocity is a frequently proposed cause of self-sacrificial cooperation in humans. Unlike *direct* reciprocity, where you remember interactions with people you yourself have cooperated with before, *indirect* reciprocity is when you recognize those who *others* have cooperated with—you gather information from other individuals' experience (Sigmund 2012).

Indirect reciprocity will work if the costly cooperation has a large enough effect on the helper's *reputation* in future interactions. This kind of explanation is a type of *costly signaling*, where the helper takes a cost to signal his or her qualities as a partner in cooperative interactions; it is an investment in reputation.

Think about how important reputation is in human interactions to see how important this kind of factor is for human cooperation. If a friend leans over and whispers 'Watch out, that guy is known to take advantage of people,' this will surely put you on your guard. And measurements indicate that about two thirds of freely forming conversation is spent on social topics—on gossip (Dunbar 2004).

To learn from others' interactions is one of several possible explanations of primarily the intricate cooperation that exists between humans. Human language seems to be a pre-requisite for such indirect reciprocity since with language we can transmit information about each other through gossip. As the Australian biologist David Haig once remarked: 'For direct reciprocity you need a face. For indirect reciprocity you need a name' (cited in Nowak 2012).

We Are Not Them: About Our Closest Relatives

Humans cooperate within families, just as many other animals, and strike up friendships, just as many other primates. Further, we learn whom we can trust by looking at others' interactions with each other. Since we are so much like other primates, does this mean that we can look at the sociality of other primates and reach insights about humans? Well yes—and no.

If the brain is important for monkeys it seems even more important in our closest relatives, the apes, whose brains on average are larger than those of monkeys. (Except that apes lack tails—is *this* the sign that we are the pinnacle of evolution?)

Of apes, gibbons are the ones we are least related to. The gibbons are a group of 13 species that inhabit the jungles of Southeast Asia. They live in life-long monogamous pairs of the type that would make religious fundamentalists dizzy with joy, and give birth, as do most primates, to one young at a time. In some species, the female and the male sing a duet together at dawn each day, perhaps to strengthen their couple's bond, perhaps to demarcate their territory, probably for both reasons.

A step closer to humans come orangutans. They live solitary lives, although females take care of their youngest offspring for years. Both males and females have territories that they defend, but males have territories that overlap those of several females. In this way, orangutans are solitary and polygynous at the same time.

Another step closer are the gorillas. They live in harems in the jungle. A dominant male holds a harem of several females and their young. The males are nearly twice the size of the females. Gorillas are thus more classically polygynous.

Our closest relatives are two species: chimpanzees and bonobos (pygmy chimpanzees). Both species live in large bands in the jungles of West Africa. Bonobos live south of the Congo River, chimpanzees on the north side. In both these groups many related males live together with a number of unrelated females in one group.

So what do the social organization of these species tell us about the social organizations of humans? Do we cooperate the way they do? If our sociality is an inheritance from our closest relations, then we should be both monogamous as gibbons and polygynous as the rest. We should live alone like orangutans, in harems like gorillas and in large groups like chimpanzees?

The problem here concerns group membership. We *are* apes and therefore for example lack a tail, but we *are not* gibbons, orangutans, gorillas, chimpanzees or bonobos. These are *other species* of apes, our relatives. Every other species of ape has its own unique solution to the specific problem of being an ape in their environment. Humans are unique among apes—just as unique as all other species are unique among apes. A look at the differences between the two species we are most closely related to should make this crystal clear.

Fifty years ago we knew almost nothing about our closest relatives. But Louis Leakey, a British anthropologist, famous for having found numerous hominoid fossils in East Africa, became interested in the behavior of apes. No one at the time knew which species was most closely related to us, so Leakey investigated all three great apes (chimpanzees, gorillas and orangutans—at the time it wasn't clear that bonobos were a separate species from chimpanzees).

Leakey went searching for someone willing to live in the jungle and spend the time it takes to habituate the animals to humans. You can't very well study animals who disappear as soon as they get a glimpse of humans, so you have to slowly, slowly make the animals used to human presence—habituate them.

In 1956 Leakey sent his secretary Rosalie Osborn to Mount Muhabura in Uganda to help with the habituation of gorillas, but she lasted only four months. Instead, a young woman named Jane Goodall became the first to study a species of ape at close quarters. She chose to study chimpanzees.

Goodall's revolutionary study of the chimpanzees of Gombe started in 1960. She came to stay for 44 years and the study she initiated is still ongoing. Goodall became the first person to habituate a group of apes and became the first person to report that chimpanzees use tools, something that made Leakey write to her excitedly '*Now we must redefine tool, redefine Man, or accept chimpanzees as humans*' (Goodall 1999) (Oddly enough, it was known already by Darwin that chimpanzees use stones to crack nuts (Darwin 1871), so it is a bit strange that Leakey was so surprised).

Goodall also made another interesting observation: chimpanzees wage war on each other. It all began when one of Goodall's chimpanzee groups split in two. Seven adult males and three females with young started to spend more and more time in the southern part of their area and eventually formed a splinter group. When males from the old group met males from the new group they would threaten each other with typical territory marking behaviors.

The conflict didn't stop at threats, and escalated. The males of the old group started patrolling the border area toward the new group. As soon as they encountered another male alone, they ganged up on him and killed him. This war-like conflict lasted for four years and didn't stop until the whole splinter group had been eradicated. The event has gone down in history as 'the four-year war of Gombe' (Goodall 1999) and is commonly used as an example of warfare in nature.

Does this means that war is inbuilt in human nature? The Army Museum in Stockholm seems to think so and open its exhibit on the history of war with a bloody installation of a group of warring chimpanzees.

However, the observation of this chimpanzee conflict does not mean that mankind is doomed to war by nature. No more than observations of chimpanzees mean that humans have hair over their whole bodies, prefer to walk on all fours and get each other's attention by using guttural noises.

You cannot derive this type of conclusion from the unpleasant behaviors of our relatives. What about conclusions from pleasant behaviors then? Let's look at our other closest relative, the bonobo.

Bonobos have now and then been described as sex-loving role models for the peaceful society. They have never been observed to have violent conflicts of the kind that is often reported with chimpanzees. Instead, all problems in bonobo societies seem to be solved by sex. Sex among all possible members of bonobo society has been observed; between males and females, between females, between males, and between young and adults.

Sex is definitely not only about reproduction in bonobos but it seems to be also used for pleasure. Bonobos have been observed to have sex face to face, to kiss each other on the mouth and to practice oral sex.

Some researchers have gone so far as to claim that bonobos use sex instead of other behaviors: to greet each other and to calm themselves, to solve conflicts and for reconciliation. Even if it may be an over-interpretation it is clear that bonobos have sex with many different types of partners, in many ways and much more often than other apes.

So are humans over-sexualized by nature because of their relatedness to this love-primate? The answer, again, is no. Just as chimpanzees and bonobos differ in their uses of sex and violence we are potentially unlike them concerning sex and violence. Humans are neither chimpanzees nor bonobos and thus we are doomed neither to exaggerated violence nor exaggerated sex (nor the opposite, of infrequent violence and vanilla sex).

Reciprocity in Humans

Humans have our own evolutionary history that differs from the other apes'. Every species of ape has its own traits—and so do we. You cannot draw any conclusions about human behavior from isolated examples of what other apes do. It is general patterns informed by evolutionary theory that can help us understand human cooperation, not isolated examples. So are there any signs of reciprocity in humans?

Are there ever!

It is easy to experimentally confirm traces of reciprocity in human psychology. Remember, in reciprocal interactions it is important to remember 'freeloaders'—those tagging along for free on the cooperation of others by receiving help but not reciprocating. These individuals should be shunned in future interactions, or reciprocity breaks down.

Psychological experiments indicate that humans have much better memories for faces than for any other patterns, and what is more, we remember the faces of freeloaders best. In experiments, researchers have shown pictures of people with different associated descriptions, and in cases where the person in the picture was described as a freeloader in social interactions the experimental subjects remembered the face longer than they remembered faces of people with nicer descriptions (Mealey et al. 1996; Oda et al. 2015).

However, there are also a number of studies that couldn't replicate these findings (Barclay and Lalumière 2006), so the jury is still out. Face recognition is of obvious importance for both direct and indirect reciprocity, so it wouldn't be strange to have an evolved special ability for this (alternatively, can be learnt). Nevertheless, one study indicates that people at least *rate* remembering faces of cheaters more important than remembering other faces—and they consequentially spend a longer time looking at them (Chiappe et al. 2004). How well they succeed is another issue.

It has also been observed that reciprocal interaction with people that you trust turns on the brain's reward system and provides a sense of bliss. If you are cooperating with people that you do not have the same trust for, this system does not turn on (Phan et al. 2010). Well-functioning cooperation makes you feel good.

There are even special neurons in the brain that turn on both when you carry out some action and when you see someone else carry out the same action, so called mirror neurons (Rizzolatti and Craighero 2004). So far, research on these neurons has not revealed any real insight as to the exact function of these neurons, but they may turn out to be very important for our ability to understand the behaviors of others and for our ability to imitate.

Nevertheless, we humans obviously have specially evolved mechanisms in our psychology for reciprocal interactions with other individuals. Direct and indirect reciprocity are thus biological explanations of human cooperation, in addition to cooperation within families.

As we have seen, evolutionary biologists reckon that a lot of cooperation is understandable through purely biological mechanisms—that cooperation between individuals 'really' is due to underlying patterns of genes that either profit in the long term from reciprocal exchanges or because they help themselves via other individuals—selfish genes.

As we saw in the last chapter, family is an important part of human sociality because of kin selection. What we have explained in this chapter is friendship, reputation and trust via direct and indirect reciprocity.

So both family and friendship are governed by our genes' 'selfish interests'? This is a deeply disconcerting thought to some. As soon as the explanation diverts from 'we help each other because it is the right thing to do'; as soon as this 'right thing' gets a definite and measurable meaning, something diminishes our humanity.

One researcher who took the problem seriously was the American population geneticist George Price. He is known chiefly for the so called Price equation, a mathematical description of evolution and natural selection, discovered when Price was trying to derive Hamilton's rule. He is also known for introducing game theory to biology.

The question whether 'real' unselfish behavior exists disturbed Price. He was deeply unsettled by Hamilton's equations that indicated that individual altruism was caused by genetic self-interest. Is altruism only an illusion? His thoughts took a drastic turn for the worse when a religious vision prompted him to study the New Testament and gave away all his belongings to help homeless alcoholics. When the people he helped stole from him he slipped into deep depression. He was ultimately evicted and became even more depressed from not being able to help anymore. Finally, he killed himself by slitting his own throat with a pair of scissors. Bill Hamilton identified his dead body (Nowak and Highfield 2011).

Such a dark depressing view of altruism as largely due to selfishness on the genetic level is unnecessary. It is possible to reverse the perspective, as philosopher Peter Singer does. Singer says that the *cause* of cooperation is uninteresting. The main point must really be the end result: that we *act* altruistically.

Even if unselfishness so far only has been explained in terms of relatives and people that we expect to help us back, those *are* still two natural mechanisms that actually *do* make people behave unselfishly towards each other.

And human cooperation *is* larger than this, because we haven't yet explained all human cooperation using kin selection and the two versions of reciprocity. Humans also cooperate with others beyond family and friends—we regularly cooperate with people that we never expect to interact with or even meet again. Why? How did that evolve? The next chapter deals with this.

References

Axelrod, R., & Hamilton, W. D. (1981). The evolution of cooperation. *Science, 211*, 1390–1396.

Barclay, P., & Lalumière, M. L. (2006). Do people differentially remember cheaters? *Human Nature, 17*, 98–113.

Binmore, K. (2007). *Playing for real: A text on game theory*. Oxford: Oxford University Press.

Brosnan, S. F., Schiff, H. C., & de Waal, F. B. M. (2005). Tolerance for inequity may increase with social closeness in chimpanzees. *Proceedings of the Royal Society, Series B: Biological Sciences, 272*, 253–258.

Brown, A. (2005). The kindness of strangers. *The Guardian* 27 August.

Bshary, R., & Noë, R. (2003). Biological markets: The ubiquitous influence of partner choice on the dynamics of cleaner fish-client reef fish interactions. In Hammerstein, P. (Ed.), *Genetic and cultural evolution of cooperation* (pp. 167–184). Dahlem Workshop Reports, London: MIT Press.

Cheney, D. L., & Seyfarth, R.M. (1980). Vocal Recognition in Free-Ranging Vervet Monkeys. *Animal Behaviour, 28*, 362–367.

Chiappe, D., Brown, A., & Dow, B. (2004). Cheaters are looked at longer and remembered better than cooperators in social exchange situations. *Evolutionary Psychology, 2*, 108–120.

Clutton-Brock, T. H. (2009). Cooperation between non-kin in animal Societies. *Nature, 462*, 51–57.

Darwin, C. (1871). *The descent of man, and selection in relation to sex* (1st ed.). London: John Murray.

Dunbar, R. I. M. (1992). Neocortex size as a constraint on group size in primates. *Journal of Human Evolution, 22*, 469–493.

Dunbar, R. I. M. (2004). Gossip in evolutionary perspective. *Review of General Psychology, 8*, 100–110.

de Waal, F. B. M. (1989). Food sharing and reciprocal obligations among chimpanzees. *Journal of Human Evolution, 18*, 433–459.

de Waal, F. B. M. (1997). Food-transfers through mesh in Brown Capuchins. *Journal of Comparative Psychology, 111*, 370–378.

Fletcher, J. A., & Zwick, M. (2006). Unifying the theories of inclusive fitness and reciprocal altruism. *The American Naturalist, 168*, 252–262.

Goodall, J. (1999). *Reason for hope: a spiritual journey*. New York: Warner Books.

Krupp, D. B., DeBruine, L. M., & Jones, B. C. (2011). Apparent health encourages reciprocity. *Evolution and Human Behavior, 32*, 198–203.

Kuhn, S. (2014). Prisoners' dilemma. In E. N. Zalta (Ed.), *The Stanford Encyclopedia of Philosophy* (Fall 2014 ed). http://plato.stanford.edu/archives/fall2014/entries/prisoner-dilemma/

Lindenfors, P., Nunn, C. L., & Barton, R. A. (2007). Primate brain architecture and selection in relation to sex. *BMC Biology, 5*, 20.

Lindenfors, P., & Tullberg, B. S. (2011). Evolutionary aspects of aggression: The importance of sexual selection. In R. Huber, D. L. Bannasch, & P. Brennan (Eds.), *Aggression: Advances in genetics* (Vol. 75, pp. 7–22). Burlington: Academic Press.

McCarty, C., Killworth, P. D., Bernard, H. R., Johnsen, E., & Shelley, G. (2001). Comparing two methods for estimating network size. *Human Organization, 60*, 28–39.

Mealey, L., Daood, C., & Krage, M. (1996). Enhanced memory for faces of cheaters. *Ethology and Sociobiology, 17*, 119–128.

Mellis, A. P., & Semmanm, D. (2010). How is human cooperation different? *Philosophical Transactions of the Royal Society, Series B: Biological Sciences, 365*, 2663–2674.

Milinski, M. (1987). TIT FOR TAT in sticklebacks and the evolution of cooperation. *Nature, 325*, 433–435.

Noë, R., & Voelkl, B. (2013). Cooperation and biological markets: The power of partner choice. In K. Sterelny, R. Joyce, B. Calcott, & B. Fraser (Ed.), *Cooperation and Its Evolution* (pp. 131–151). MIT Press: Cambridge.

Nowak, M. A. (2006). Five rules for the evolution of cooperation. *Science, 314*, 1560–1563.

Nowak, M. A. (2012). Evolving cooperation. *Journal of Theoretical Biology, 299*, 1–8.

Nowak, M. A., & Highfield, R. (2011). *Supercooperators*. London: Canon Gate.

Oda, R., Kato, Y., & Hiraishi, K. (2015). The watching-eye effect on prosocial lying. *Evolutionary Psychology, 13*, 1–5.

Phan, K. L., Sripada, C. S., Angstadt, M., & McCabe, K. (2010). Reputation for reciprocity engages the brain reward center. *Proceedings of the National Academy of Sciences, 29*, 13099–13104.

Queller, D. C. (1985). Kinship, reciprocity and synergism in the evolution of social behaviour. *Nature, 318*, 366–367.

Read, L. E. (1958). *I, Pencil: My family tree. As told to Leonard E. Read.* http://www.econlib.org/library/Essays/rdPncl1.html

Rizzolatti, G., & Craighero, L. (2004). The mirror-neuron system. *Annual Review of Neuroscience, 27*, 169–192.

Sigmund, K. (2012). Moral assessment in indirect reciprocity. *Journal of Theoretical Biology, 299*, 25–30.

Silk, J. B. (2013). Reciprocal altruism. *Current Biology, 23*, R827–R828.

Silk, J. B., Beehner, J. C., Bergman, T. J., Crockford, C., Engh, A. L., Moscovice, L. R., et al. (2010). Strong and consistent social bonds enhance the longevity of female baboons. *Current Biology, 20*, 1359–1361.

Tomasello, M. (2009). *Why we cooperate*. Boston: Boston Review.

Tooby, J., & Cosmides, L. (1995). The psychological foundations of culture. In J. H. Barkow, L. Cosmides, & J. Tooby (Eds.), *The adapted mind: Evolutionary psychology and the generation of culture* (pp. 19–136). Oxford: Oxford University Press.

Trivers, R. L. (1971). The evolution of reciprocal altruism. *Quarterly Review of Biology, 46*, 35–57.

Trivers, R. L. (1972). Parental investment and sexual selection. In B. Campbell (Ed.), *Sexual selection and the descent of man, 1871–1971* (pp. 136–170). Chicago: Aldine.

Trivers, R. L. (1974). Parent-offspring conflict. *American Zoologist, 14*, 249–264.

Trivers, R. L. (1991). Deceit and self-deception: The relationship between communication and consciousness. In M. Robinson & L. Tiger (Eds.), *Man and beast revisited* (pp. 175–191). Washington, D.C.: Smithsonian.

Trivers, R. L., & Hare, H. (1976). Haplodiploidy and the evolution of the social insects. *Science, 191*, 250–263.

Trivers, R. L., & Willard, D. E. (1973). Natural selection of parental ability to vary the sex ratio of offspring. *Science, 179*, 90–92.

TT/The Local. (2007). *Swedish tax collectors organized by apes.* https://www.thelocal.se/20070723/7972

Van Schaik, C. P., & Kappeler, P. M. (2006). Introduction. In P. M. Kappeler & C. P. van Schaik (Eds.), *Cooperation in primates and humans* (pp. 3–21). Berlin: Springer.

Virtanen, J. (2005). *Robert Trivers on Deceit, self-deception, and warfare.* http://www.hotales.org/writings/trivers-on-deceit-self-deception-and-warfare.html

Wynne, C. (2007). Aping language: A skeptical analysis of the evidence for nonhuman primate language. *eSceptic.* http://www.skeptic.com/eskeptic/07-10-31/#feature

West, S. A., El Mouden, C., & Gardner, A. (2011). Sixteen common misconceptions about the evolution of cooperation in humans. *Evolution and Human Behavior, 32*, 231–262.

Wilkinson, G. S. (1990). Reciprocal food sharing in the vampire bat. *Nature, 308*, 181–184.

Humanity—The Paragon of Cooperation?

Abstract This is the crucial chapter upon which the book turns, outlining the problem with human cooperation; that it goes over and above the biological explanations of cooperation that we have discussed so far. The theory of strong reciprocity is presented, as well as many game theoretical experiments, their results and implications. Cultural group selection is discussed and the idea of cultural inheritance is introduced.

> What a piece of work is a man! How noble in reason! How infinite in faculty! In form and moving how express and admirable! In action how like an angel! In apprehension how like a god! The beauty of the world! The paragon of animals!
>
> From Shakespeare's Hamlet, written ca. 1600.

Human cooperation is special. We regularly cooperate with people we are not related to and do not know. Cooperation between humans occurs within very large groups, with individuals we meet only fleetingly and perhaps never again.

As we have seen, kin selection helps explain cooperation with related individuals, and direct and indirect reciprocity help explain cooperation in small groups where repeated interactions occur and you can know each person's reputation. But what about groups as large as cities or countries, where routine interactions and cooperation occurs between people who are unrelated, unacquainted and may never meet again?

Some use the term *strong reciprocity* for this human proclivity for cooperating with one-off acquaintances and the related tendency to pay to punish one-off 'freeloaders' (Fehr et al. 2002). But what does 'reciprocity' imply in the term 'strong reciprocity'? (Burnham and Johnson 2005) 'Reciprocity' normally refers to repeated interactions; but here it (often) alludes instead to single acts of selfless behavior towards individuals one has never met. Though inappropriate, the expression is an accepted one, at least currently, although I will instead be using the term 'exaggerated cooperation' or similar.

One early view of human cooperation is that societies arise as by-products of the selfish actions of many—and now we're talking about selfish individuals, not selfish genes. This view was best articulated by Adam Smith in 1776 in *The Wealth of Nations*. Regarding the selfish person, Smith writes:

P. Lindenfors, *For Whose Benefit?*, DOI 10.1007/978-3-319-50874-0_7

> … [H]e intends only his own gain, and he is in this, as in many other cases, led by an invisible hand to promote an end which was no part of his intention. Nor is it always the worse for the society that it was no part of it. By pursuing his own interest he frequently promotes that of the society more effectually than when he really intends to promote it. (Smith 1904)

But if we are all inherently selfish, how is it that many of us, in interactions with complete strangers, 'give' something? Small voluntary acts intended to benefit others are so common in human societies that they have been given their own term by social scientists: 'prosocial behaviors'. Why do we give money to beggars, hold up doors for others, apologize, turn in things to the Lost and Found, and perform other trivial acts of generosity that characterize a civilized society?

And, even harder to explain, how come some people risk their life and limb to defend others they do not know? Why are there policemen and soldiers? Why would anyone want to sacrifice their life for something as abstract as 'country', 'justice', 'God' or 'freedom'?

The problem with this kind of unselfishness is aptly illustrated in a Slovak saying: 'He who does not steal, steals from his family' (Rankin 2011). Helping others in a costly or risky way rebounds on those one should be cooperating most with. How can such costly cooperation exist?

Further, in large communities, anonymity is easy. Large groups are vulnerable to individuals who cruise through and take advantage of naïve cooperators (Enquist and Leimar 1993). Why don't transient freeloaders break societies?

One school of thought in anthropology and sociology—functionalism—holds that norms, rules and social institutions exist to regulate societies and create a workable, well-functioning unity. If we understand how all these components work for *the good of the group*, we can understand how they have arisen and thereby understand the unit—society. In functionalism, individuals are only cogs in a machine. It is no coincidence that this sounds like a model based on the idea that society works like a living organism; many concepts derive from biology (Laland and Brown 2002).

Because there are genetic explanations for all other cases of cooperation between individuals in nature, many biologists are open to a genetic cause for mankind's exaggerated cooperation too. Characteristics common to all individuals of a species generally have a genetic background and we all seem to cooperate 'too much'. What could be the genetic logic here? How would a 'selfish gene' benefit from cooperation outside of family and acquaintances?

Let us re-affirm that people cooperate for reasons already reviewed: cooperation delivers direct advantages; it favors 'genes for cooperation' in related individuals; we regularly exchange services; it teaches us whom are to be trusted when we observe others exchanging services, or listen to their gossip.

We are now faced with explaining a form of cooperation that is beyond this, something that exists only among humans: taking a cost—from the trivial (holding doors open) to the fatal (giving one's life for one's country)—towards completely unknown individuals and without predictable return. In such interactions there appears to be only loss, no gain. In many cases giver and recipient will never meet again.

The will to cooperate with strangers seems to exist in humans from a very early age. Drop something on purpose in front of a one-and-a-half-year-old and, most often, the child will pick it up and hand it to you, even if you are a stranger. Equally, if you ask an unfamiliar child to pass you something or open a door, the child will generally comply.

Experiments have shown that juvenile chimpanzees act similarly (though you have to arrange the experiment a bit differently). Uniquely, however, human children will point to things in order to help; that is, they will act, using body movement, simply to convey information. Chimpanzees never provide information in this manner. Children will also acquiesce to simple arbitrary rules, for example in games. If an outsider arrives and attempts to play a game, but in a different way, children will become aggrieved. An agreement is being breached. 'No, not that way!' the child will protest.

Children are fond of mimicking the behavior of others, even self-sacrificing behavior. To show this, researchers have had children play games with tokens as a reward. The tokens can be exchanged for toys, with better toys costing more tokens. The trick the testers employ is to stage the test area with a collection box. If an adult demonstrates the game and then donates the tokens to the collection box, the tested children will do so more often than when the adult keeps the tokens. This holds true even when the children are left alone in the room (Henrich and Henrich 2007).

A professor of psychology at the Max Planck Institute in Leipzig, Michael Tomasello, has specialized in this type of test involving children and chimpanzees, to examine to what extent our helpful attitude is innate from childhood and whether or not it is unique to humans. He has found that to a large degree, helpfulness and cooperation seem to exist from an early age in child development.

Together with co-workers, Tomasello have subjected chimpanzees, orangutans and 2½-year old German children to 38 cognitive tests and remarkably found almost no differences in mental abilities that concern the physical world between chimps and children. There were, however, incredible differences concerning social learning subtests, where children scored almost perfectly while the apes were completely unable to accomplish the tasks. If we were to give such tests to adult humans, we would score perfect every time. Not so with chimps, where adults and juveniles are equally incompetent.

Tomasello further believes that humans are expert in quickly establishing a feeling of 'us' and common goals. This can also be established with chimpanzees, to a degree. If a human and a chimpanzee need to work together to solve a problem, the chimpanzee will cooperate. But if the human abandons the effort, the chimpanzee will follow suit, not attempting to coerce the human to continue.

Children, on the other hand, will *insist* on continuing. If the activity is set up to give the child its reward before its partner, the child will want to continue until both have been rewarded. Not so with chimpanzees who just stop. Tomasello suggests that this 'extra' human cooperation between strangers is the result of a frequent, particularly human, need to resolve situations by common goals and common rewards (Tomasello 2009; Warneken 2013 and references cited therein). If Tomasello is right, exaggerated human cooperation is a by-product of mutually beneficial cooperation.

We saw another theory in the previous chapter: that human language brings with it the potential of gossip (Enquist and Leimar 1993). In any community, you are permanently under scrutiny; people see what you do and how you act. If you demonstrate good social behavior you will be seen as polite and friendly and your good reputation will spread thorough the community. You are then free to behave badly in private since the basis of your social status has more to do with the maintenance of a façade of good behavior in public. People are amazingly good at keeping tabs on others' status (Mellis and Semmanm 2010).

This is indirect reciprocity but with a reinforced effect since we can talk to each other and thereby communicate remote information about the trustworthiness of others. If others are always watching and one feels constantly observed, the possibility exists of demonstrating generous behavior towards strangers as a signal of personal quality (Zahavi and Zahavi 1997).

The Israeli evolutionary biologists Avishag and Amotz Zahavi have suggested that the explanation for exaggerated cooperation is just that, that we have a constant need to demonstrate our cooperative quality. Indirect reciprocity allows human societies to be larger than otherwise possible. People can for example establish reputations for being good at specific jobs. Because achieving such a good reputation for specialized tasks, communities can even sustain people who write books about cooperation while others grow food.

Indirect reciprocity facilitates specialization, one of the central components of synergetic cooperation—cooperation in which the whole is greater than the sum of its parts. Our societies are so much more than merely groups of individuals cooperating with family and friends.

But this ushers in a new problem. In a large community such as a city, it is possible to disappear in the crowd. An individual can thrive by repeatedly exploiting those who trust their fellows or who signal their trustworthiness by beginning interaction by giving favors or money (Enquist and Leimar 1993). If you think about it, you don't really know all the individuals you cooperate with. The bus driver? The check-out clerk in the convenience store? The saleswoman? The bank teller? There are plenty of opportunities to take advantage of naïve cooperators.

Direct and indirect reciprocity suffice to explain cooperation with people we repeatedly interact with or whose reputations we are aware of. But how do we explain that we regularly cooperate with strangers as well? To find an answer to this puzzle of human cooperation—the very foundation of society—we need to delve deeper into this issue.

Games of Cooperation

An interesting illustration of how mankind's 'extra' cooperation works is illustrated in a simple experiment called *the ultimatum game.*

Two players who don't know each other are given a sum of money. One of the players, A, decides unilaterally how the money is to be divided. When A makes an offer (the ultimatum), the other player, B, can choose between two things: accepting the money offered, or refusing it. If B refuses, neither player gets any reward. If B accepts, it will be according to the division proposed by A. The players are anonymous to each other—they interact via computer screens—and are told in advance that their identities will never be revealed, and nor will they play against each other ever again.

If people were strictly rational and counted only gains, B would accept *any* offer from A; it doesn't matter to B how much A stands to gain, any gain for B is a gain. For a neoclassically trained economist, the solution to B's dilemma is thus obvious: take the money and be satisfied, whatever the share. Neoclassically trained economists who have learnt this, actually often do make this 'right' choice; they take what is offered and walk away (Frank et al. 1993).

Volunteers lacking education in neoclassical economics—that is, most of us—act quite differently. People in B's situation regularly refuse what they are offered if it is perceived as stingy or unfair. Thus, people put in player B's situation are often willing to sacrifice a sum of money just to deny player A—the money-grubber—his or her gain. People will refuse to cooperate with someone seen to be acting unjustly, even if this has a cost. Why?

If the experiment is devised cleverly enough, it can be carried out on chimpanzees (except that a chimpanzee needs to see its gaming partner to understand that it is interacting with another individual). Chimpanzees react 'rationally', like economists, taking the grapes on offer, however few they are (Jensen et al. 2007).

The obvious joke is that only chimpanzees and economists behave according to economic theory. But it was actually economists who discovered and started investigating this problem. Researchers *love* problems. And not all economists see people as entirely rational beings. The Indian Nobel Prize winner in Economy, Amartya Sen, for example, has called the implied shortsighted, selfish person in economic theory 'the rational fool' (Ridley 1996).

To try to understand what was going on, researchers constructed an even simpler game, called *the dictator game.* Again, two players are given a sum of money. But now, one of the players is appointed dictator. The dictator decides how the money is to be divided, while the second player decides nothing, not even whether to accept or reject the offer. The players are again anonymous to each other and told in advance that their identities will never be revealed to each other, and neither will they play against each other again.

Does the dictator take all the money and leave, happy over his easy good fortune? Does other player realize there is nothing left and go home empty handed but scarcely surprised? If so, the dictator would be acting rationally and the researchers could sleep easily in the secure knowledge that the predictable happened.

But that is not what occurs. The dictator almost always shares, despite the absence of possible punishment for unjust behavior, and it is not uncommon that the dictator gives away half of the money. Why?

Simple experiments like this can rock the boat for science since neoclassic economic theory is built on the premise that people are economically rational. This type of experiment [there are many more games of different types (Shermer 2008)] has generated a new field of science: *behavioral economics*.

This discipline studies the social, cognitive and emotional reasons for people's economic decisions, in contrast to the neoclassical school of economic theory that focuses instead on the way markets function. The problem with neoclassical economics is that its models are based on human psychological assumptions that are incorrect, namely that people act rationally in economic decisions, attempting to maximize their gains [they are 'rational fools' (Ridley 1996)]. As we have seen, simple experiments show that people act differently.

It seems from experiments such as those just described that fairness is something we are willing to pay for, either through sharing or through paying for punishment. It does not appear that we humans are trying to maximize gain—at least not the monetary kind—not even when we have near perfect access to information in the relevant situations, which we almost never do.

This makes neoclassic economics a discipline that analyses aspects of human cooperation where parts of the theory build on incorrect assumptions of how people are. How do economists react to this insight?

Someone who tackled the issue early was the 1978 Nobel Prize winner in Economics, American Herbert Simon.[1] Simon coined the term 'bounded rationality', which means that people's ability to act rationally is limited by variable access to information, our restricted intellect and the time pressure we often face when making decisions (Simon 1957).

This bounded rationality, believes Simon, should be incorporated in economic models based on human behavior. Another Nobel Prize winner in Economics (2009), Elinor Ostrom, agrees, stating that even if pure rationality models are still useful, a new type of model that includes knowledge of human psychology is needed (Ostrom 1998).

A great step in this direction was taken by the Israeli-American psychologists Daniel Kahneman and Amos Tversky in their prospect theory. Prospect theory is based on empirical observations of how people actually make decisions and describes how they make choices in situations involving risk (Kahneman and Tversky 1979). Kahneman was given the 2002 Nobel Economy Prize for his work

[1]The Sveriges Riksbank Prize in Economic Sciences in Memory of Alfred Nobel 1978 was awarded to Herbert A. Simon 'for his pioneering research into the decision-making process within economic organizations'.

while Tversky unfortunately died before the prize decision was made.[2] You may know Kahneman from his brilliant book 'Thinking Fast and Slow' which outlines his research on these aspects of human psychology (Kahneman 2011).

The neoclassisist economist Milton Friedman (himself a Nobel Economy Prize winner in 1976[3]) had, even before the discovery of behavioral economics, shrugged off this type of reasoning: 'The relevant question is not if the 'assumptions' backing a theory are 'realistic' because they never are, but if they are approximations adequate for the purpose. And this question can be answered by seeing if the theory works, which means that it can supply adequately correct predictions' (Friedman 1953). Critics of behavioral economics argue that there are almost no signs that our non-optimal choices in various risk situations result in non-optimal economic consequences (Berg and Gigerenzer 2010).

But even if the neoclassicist models work, as Friedman claims, it is reasonable to expect that models based on truer assumptions will work even better, as Ostrom asserts. This is an empirical question, with the answer yet to be found in the future. The challenge for behavioral economists lies in constructing models on firmer psychological base assumptions to see if these better describe how markets and human societies work.

A Huge Mistake?

If you scrutinize human communities, they are practically miracles of cooperation. Just think of how many individuals need to be involved to enable you to enjoy a ham sandwich, from raw materials to kitchen refrigerator.

Grain for the bread has to be grown, harvested, milled and baked, calling for fertilizer, tractors and other agricultural tools and machinery to be produced in several factories, these in turn demanding metals and plastic. Pigs must be raised and fed, slaughtered, butchered, the meat cured, sliced, packaged and transported, which again necessitates agricultural and factory machinery. Everything is then transported to your supermarket by vehicles running on fuel that in turn has to be extracted and refined. Each machine requires its own manufacturing factory, running on energy that itself demands a whole industry to manufacture demanding ever more machines and energy. It can seem like all of society is required just to make a ham sandwich.

---

[2]The Sveriges Riksbank Prize in Economic Sciences in Memory of Alfred Nobel 2002 was divided equally between Daniel Kahneman 'for having integrated insights from psychological research into economic science, especially concerning human judgment and decision-making under uncertainty' and Vernon L. Smith 'for having established laboratory experiments as a tool in empirical economic analysis, especially in the study of alternative market mechanisms'.

[3]The Sveriges Riksbank Prize in Economic Sciences in Memory of Alfred Nobel 1976 was awarded to Milton Friedman 'for his achievements in the fields of consumption analysis, monetary history and theory and for his demonstration of the complexity of stabilization policy'.

It has long been recognized that human societies and human technology development build on mutual dependency. In 1958, an American economist, Leonard Read, published an essay entitled 'I, Pencil', purportedly the autobiography of a lead pencil, implying that a pencil is not merely the product of an enormously complex system of sourcing and manufacturing processes (for example, sourcing wood from forests, graphite from mines and the production itself in a pencil factory). In addition, no single person commands all the skills for making a pencil, not even the manager of the pencil factory, since no individual understands all the processes involved in all the production stages (Read 1958).

Societies this complicated can obviously not arise complete all at once. At the dawn of mankind, we hunted and gathered for our 'ham sandwiches' ourselves. Since people once lived in small groups where everyone was either a relative or familiar, no cost was incurred by cooperating 'too much'. As we have seen in previous chapters, cooperation between relatives and acquaintances is something we understand rather well.

Later, with the development of agriculture, people needed to produce all the sandwich's ingredients themselves or get them from someone they had a personal relationship with. You would visit the blacksmith for tools, the miller to have your grain milled, and the butcher to slaughter and butcher your pig, and so on.

These interactions worked through reciprocal exchanges of goods and services. There was no new cooperation problem to explain. Adam Smith expressed it clearly: 'It is not from the benevolence of the butcher, the brewer or the baker that we expect our dinner, but from their regard to their own interest' (Smith 1904). It is only in recent times that interactions have become impersonal in a way that we can no longer understand. We do not any longer know where the ham and bread come from, nor how they were produced.

Reciprocal barter—trade—is an ancient human occupation. Stone axes and other artifacts from the Stone Age have been found by archaeologists far from their origins: a clear indication of trade. It is claimed by some that the first indications of such trade can be traced to Ethiopia from finds that are 80,000–120,000 years old. Jasper and obsidian have been found many tens of kilometers from their origin (Ridley 2010). This makes trade much older than agriculture, which is an invention 'only' about 12,000 years old.

Intricate trade relations are practiced also by today's surviving hunters-and-gatherers, so trade does not demand complicated organized societies. It is not either some novel form of cooperation we are talking about, but immediate reciprocity. 'You give me this and I'll give you that.'

Thus, since gains are immediate the occurrence of trade is not so remarkable. As the British economist David Ricardo noted as early as 1817 (Ricardo 1817), it is even possible to profit from trade even though you might be better at producing *everything* than the person you are trading with.

For example, imagine that caveman Ugh takes 4 h to make a spear and 3 h to make an axe (a total of 7 h). Zogh takes only 1 h to fashion a spear and 2 h to produce an axe (a total of 3 h). If Zogh makes 2 spears (2 h) and Ugh makes 2 axes (6 h) and they trade a spear for an axe, both have saved an hour's work, even though Zog can do everything Ugh can, but better.

Barter trade is thus profitable even at a very basic level. Yet there is no evidence of bartering among Neanderthals; it seems to be an ability specifically belonging to modern humans.

Traditional trade is often carried out between groups thoroughly familiar with each other. Only in recent times have interactions become impersonal to the extent that we no longer really understand them.

Perhaps exaggerated cooperation in modern societies is a kind of emergent property from behavior stemming from earlier evolutionary adaptations—from scaled up trade relationships? When we lived in small groups where we were either related to or acquainted with everyone else, or cooperated through immediate reciprocation, it may have been profitable to cooperate with *everyone*.

The problem is that a rule enjoining 'cooperation with everyone' is never going to be profitable. If there is only a single freeloader in the group, he or she must be identified; otherwise the freeloader gets a disproportionate evolutionary advantage. And if being a freeloader is advantageous it will quickly become common to the cost of those who cooperate. This is how natural selection works.

But perhaps we think that everybody else is a relative or a friend? There are researchers who see this as a solution to the problem: that our exaggerated cooperation is simply based on a gigantic mistake. However, this is impossible for the same reason: such costly mistakes would result in the rapid increase in numbers of freeloaders.

Can our large brains help us keep track of multiple social interactions? Several researchers, myself included, have shown that the degree of sociality among apes is linked to the size of their neocortex, the thinking section of the brain (Lindenfors et al. 2007). We humans have bigger brains than apes, so shouldn't we be able to cooperate with more individuals?

This explanation isn't sufficient either. As we have noted, human societies involve a large number of interactions with individuals destined never again to meet—this is the cooperation that needs to be explained. A large brain is no help when it comes to judging isolated interactions with strangers—there is nothing to remember or evaluate.

Cultural Group Selection

> Doubtless, a tribe with many members driven by strong patriotism, devotion, dutifulness, courage and compassion—and therefore always ready to help each other and to sacrifice for the common good—would defeat most other tribes. This would be natural selection.
>
> Charles Darwin: The Descent of Man, and Selection in Relation to Sex, 1871.

There is another vital human capacity that we now need to consider: we transfer information between generations by means other than genes. We have culture.

If you're not a natural scientist, you're probably shaking your head in disbelief at this point. Of course people have culture! Isn't that the premise of almost everything human?

Actually, many of our attributes have a biological background and are shared with other animals, including our tendency towards cooperation with relatives and friends. Attempts have been made to explain even our exaggerated cooperation as a biological mechanism, but it seems that biology, in this instance, is insufficient.

A clarification: I am here using the word 'culture' as it is used in research on cultural evolution: to describe traits that can be transferred between generations by other means than genetic heritage. Cultural traits in this sense are therefore traits that can be learned, but not those handed down through genes.

A possible explanation for exaggerated human cooperation, based on knowledge of social inheritance, is that our ability to learn from each other produces what is called cultural group selection. In human development, groups of cooperating individuals seem to have had an evolutionary advantage over groups of selfish individuals, which in extension can have led to a genetic alteration of our propensity to cooperate.

Non-biologists often think that qualities that are good for groups but bad for individuals are common in nature. You often hear in nature programs, 'because it's good for the species' as an explanation. But as we saw earlier, group selection is immensely rare in nature and works only under very specific circumstances. It's easy to see why. Imagine a trait good for the group but bad for the individual: for example, refraining from having children when resources are scarce. In the next generation, which trait would be more common: refraining from having children or having children? Even if the *population* would do well if everyone sacrificed a little, the *individual's* prime problem is competition with neighbors. Qualities that favor the individual (almost) always trump qualities that benefit the group.

Researchers who study cultural group selection believe, however, that it is by learning from each other that tribes are formed, as well as other groups of individuals who share cultural qualities such as language and rituals. We learn most from those closest to us.

Since we learn from each other, a common language evolves and can give rise to a common view of ethics, morals and other culture-specific behavior that ultimately identifies and directs the group—*group norms*. There has recently been an enormous interest in the evolution of such normative cognition (Hauser 2006; Kitcher 2011; Henrich 2016). Helped by nature, man can thusly have constructed an evolutionary environment of his own. This has subsequently impacted our genes, giving us a genetic explanation for our special way of cooperating, even though the background may be cultural.

Nature or Nurture

A part of the theory of cultural group selection is that there must exist a control mechanism to manage freeloaders—the group must, in some way, punish those who will not cooperate or else group coordination collapses. Does such punishment exist and does it help to maintain extra human cooperation?

Two Swiss economists, Ernst Fehr and Simon Gächter, put this to a simple test. They let 240 volunteers play another cooperation game, a so-called commons game (Fehr and Gächter 2002). Four people play at a time; none can see the others, and no two play against each other more than once.

An almost identical experiment was conducted as early as 1986 by the Japanese sociologist Yamagishi (1986), but I have chosen to describe Fehr and Gächter's version because the construction is simpler and easier to understand. But similar results were already published when Fehr and Gächter carried out their study—the one that became famous. (For some reason, Fehr and Gächter do not reference Yamagishi—a kind of omission frowned on by researcher peers.)

Each volunteer in the experiment gets 20 units to invest in a communal pot or to keep. For each unit invested, everyone including the investor gets a dividend of 0.4 of a unit. This means that if no one invests, everyone ends the game with 20 units, but if everyone invests everything they have, each player will end up with 32 units (20 units × 4 persons × 0.4 unit).

However, within the construct of the game, it is *always* profitable to sit on one's units, regardless of what the others do. If you kept your units while all others cooperate, you get your dividend regardless and still keep all your units: a total of 44 units (20 units × 3 persons × 0.4 unit + your own 20). But if everyone refrains from investing, everyone walks away with only 20 units each.

As the games continue in new constellations, investments in the common pot drop. In the first round, most players invest about half their units. But someone always succumbs to the temptation to make a little extra by investing a little less. In the next round, many choose a more careful strategy, wise from experience. But once again, someone always tries to keep a little extra. In the subsequent games, investments dive and keep sinking. At the end, more than 75% of the players are investing less than 5 units in the pot.

The game is a variation for several players of the Prisoners' Dilemma. But in a multi-player game, the tit-for-tat strategy does not work. If you cease cooperating to punish a freeloader—by not investing in the pot—you punish not only that player but the whole group. The others, unjustly punished, can easily assume that the freeloader is you, since there is no difference between not investing and punishment.

The game aptly illustrates what is usually called the Tragedy of the Commons. This is a dilemma named for a famous article in 1968 by the American ecologist Garrett Hardin in which he pointed out that individuals acting in their own interests will over-exploit or under-invest in common resources, even when everyone realizes that this is in no one's interest. Hardin came to the conclusion that the use of common resources had to be regulated by the state or by international agreements

(Hardin 1968). The game of commons is an experiment to investigate the tragedy of the common (Rankin et al. 2007).

Fehr and Gächter went a step further. In an added twist, each player was given an extra turn at the end of the game: everyone could now pay a sum of units to punish one of the others: the presumptive guilty party was fined 3 units for every unit others paid. Players were eager to punish freeloaders. Even though they would never play against the others again, and even though everyone was anonymous, punishment was inflicted with glee. (In his version, Yamagishi let the players contribute to a 'punishment pot' used to hurt the least willing investor.)

The introduction of punishment had its effect. In subsequent rounds, investments to the common pot rose instead of sinking. Ultimately, more than 75% of the players invested more than 15 units and four of ten invested *all* their units.

Yamagishi, and later Fehr and Gächter, showed that people will pay for the opportunity to penalize freeloaders and that this, in turn, prompts functioning cooperation in groups of anonymous players. Does this imply that the existence of sanctions explains people's exaggerated cooperation?

Not really, since this explanation has its own problem. It is now possible to be a freeloader in a new way: by never paying the price of penalizing someone. Especially when interactions are repeated, it does not pay to penalize (Dreber et al. 2008).

Dodging payment means you get the advantage of cooperation while avoiding not only punishment but also the cost of penalizing others. Practitioners of this new way of avoiding cost are called *second-order freeloaders*. The game-creators solved one problem but created another.

Oddly, the explanations for how second-order freeloaders are kept in check are the same as originally used to explain man's unique desire to cooperate: cultural group selection and the importance of maintaining a good reputation (Fehr 2004; Panchanathan and Boyd 2004). Groups that include cohesive penalizes perform better than groups that include freeloaders. And perhaps refraining from punishing corrodes a good reputation?

Supporters of this interpretation of the altruistic punishment theory say there is an important distinction. A mathematical asymmetry exists between the two ways of being a freeloader: since first-order freeloading means not paying to the common pot, the difference between freeloaders and cooperating individuals in a group is only an individual investment. Regardless of how many freeloaders there are in a population, the difference is always the investment in the pot and only that.

When, on the other hand, payments are made to punish freeloaders, it is expensive to be a punisher if the group includes many freeloaders. With few freeloaders it is cheap. Since the number of freeloaders differs between groups—variation is always present—the advantage of being in a group with many cooperating individuals can be greater than the small cost of penalizing isolated freeloaders. This is a group selection argument, with a cultural twist (Boyd et al. 2003).

This explanation works especially well if the evolutionary growth of altruistic punishment occurs in small groups of a few tens of closely related individuals. Rather as people lived during the early stages of our evolutionary history …

... but in those groups we have already attributed cooperation to kin selection and reciprocity!? Plus, this type of penalization cannot explain the results of the dictator game, where people give away units without risk of punishment? (Burnham and Johnson 2005).

And there are more problems. If two rounds with penalties are allowed, those who have been penalized start reciprocating in turn. If opportunity is provided, about a quarter of all penalties lead to payback on the penalizer and cooperation breaks down once again (Nikiforakis 2008). The solution seems to be to allow the group to agree on a punishment strategy to be applied automatically, as in Yamagishi's original experiment (Ostrom et al. 1992). This way, no player can extract revenge on any other individual.

If you think about it, this is how modern societies work—we do not fine freeloaders ourselves, but have organized penalizing systems through police and courts. If players are allowed to communicate and agree on penalties, however, interaction is no longer anonymous and the rules have been changed. We will return to this in the next chapter.

This research is as yet incomplete. We simply do not know if this approach can solve the mystery of exaggerated human cooperation. In the meantime, some researchers promote another way of looking at the problem completely. They're asking whether genetics are useful at all to explain mankind's exaggerated cooperation.

Cultural Explanations for Extreme Cooperation

If anything makes mankind unique, it is, as we have seen, the existence of an advanced culture. Can cooperation be explained *uniquely* in cultural terms? If so, cooperation should differ between cultures. Do some cultures cooperate more than others—are there different *norms* about cooperation in different societies?

There is some evidence of a genetic explanation of differences in cooperation. Scientists have found a genetic component in the degree of acceptance of costs that promote and maintain cooperation. By having genetically identical twins play the dictator game then comparing investments when non-identical twins play, the genetic component could be shown to represent about 20% (Cesarini et al. 2009). But despite this genetic predisposition, a cultural factor may also figure in the background.

To test this, researchers set up games in different circumstances. Volunteers would, for example, play the ultimatum game in virtual cyber communities using avatars (graphical representations of a user's character). When a player uses a tall avatar, she keeps more money when playing as player A (who is offering the ultimatum) and is less willing to accept an unfair division of the sum when playing as player B (who is appraising the offer). When playing with a short avatar, the opposite is true. The players are thus inclined to cooperate differently depending on whether they envision themselves as smaller as or larger than an opposing player (Nee and Bailenson 2007).

It also seems that people carry out the same transactions but for different reasons depending on the game being played. Interviewed after having shared money in the ultimatum game, about half the players say they did so for fear of not getting anything. They were weighing up their chances of making a profit. Asked the same question, players in the dictator game say they shared money units because to do so was fair (Henrich and Henrich 2007). Same transaction, different motivation.

The specific situation also seems to be crucial for how the game is played. Normally in the ultimatum game, the player identities (player A and player B) are assigned randomly. But when a simple quiz is used to allocate identities, something interesting happens. In this situation, player A will offer less than normal and player B will accept much smaller sums than normal. The results of the quiz have established a hierarchy in which both players feel that whoever has answered more correctly deserves more money (Hoffman et al. 1994).

Influencing people's behavior is even easier than that. Pictures of eyes on the wall—an abstract representation that someone may be watching—make volunteers less inclined to act selfishly (Nowak and Highfield 2011; Oda et al. 2015). If the experimental subjects have to make their decision quickly they behave more generously than if they are given time to think things over (Rand et al. 2012). Also, giving the game a different name alters the inclination to cooperate; if the game is called the 'Wall Street game', players will be selfish, but if it's named the 'Community Game' they will be far more willing to cooperate (Liberman et al. 2004). Extrapolating from this, the view people have of what situation they are in strongly influences their behavior. This can be demonstrated even in real life.

In a famous experiment, several day care centers in Israel imposed fines on parents who collected their children late. The effect was not the desired one: instead of the parents feeling bad about coming late, they discovered they could pay for late collection—and felt it was worth it. Their view of what type of interaction they were partaking in had changed. Instead of having a bad conscience, the parents now felt they were involved in a business transaction. The fine was discontinued but it didn't help—the image of the interaction had lastingly changed (Gneezy and Rustichini 2000).

With reference to results like these (Cardenas et al. 2000), some scientists are calling on politicians to exercise caution in using fees and rewards to prompt people to behave in ways deemed desirable for society (Bowles 2011). To assume that people are basically egotistical and to design measures based on this assumption can lead people to alter their conception of what kind of interactions they are engaged in and thereby actually create rational egoists (demonstrated with utmost clarity by the day care experiment) is not always the desired result.

Do these results indicate that societies differ on how much they cooperate? Do different societies imply different situations for people—do norms around cooperation differ? To investigate the possible effects of cultural factors, scientists have travelled widely and tested the games on various ethnic groups. And yes, differences did show up.

In some cultures, players invest only small sums in the dictator and ultimatum games, and even less in the commons games. For example, players from among the Hadza (hunter-gatherers from northern Tanzania) often refrained from offering anything at all in the ultimatum game, offering on average 27% of the sum. Opposing players experienced this as unjust and frequently refused.

Other cultures were willing to part with much larger sums. For the Lamalera, a traditional whale-hunting people in Indonesia, their catch is hard to monopolize, whales being the size they are. The profits are huge but the advantages of selfishness negligible. Here, players regularly invested a large proportion of their personal sums. The average investment was 58%.

The researchers also found interesting ways of playing the games among the Au and Gnau peoples of New Guinea, two ethnic groups that, despite similar names, speak completely different languages. Here, some players at times could offer more than half their personal sums to the other player but the ultimatum would still be rejected. Among these ethnic groups, as with several others in Melanesia, an accepted gift obligates the receiver to provide support in conflicts or respond with an equally fine gift—the finer the original gift the greater the obligation.

But there are further complications. Some societies exhibit what are called antisocial punishments: players punish others for investing *too much* in the commons game's pot. Researchers who discovered the phenomenon found it most frequent in societies where people have little trust in institutional justice and where cheating on taxes is regarded as a trivial offence (most noticeable in countries such as Oman and Greece, while the USA and Denmark are found at the other end of the scale) (Herrmann et al. 2008; Gächter et al. 2010). The existence of punishments does not, therefore, always facilitate cooperation.

So there are differences in cooperative norms between ethnic groups, and the differences are even systematic. It has been claimed that an important causal factor is the degree of integration in the market economy. The more familiar people are with market interactions, the more they will give to their opponent (Henrich et al. 2005) which may explain why upper-class people are measurably more pro-social than lower-class people across many cultures (Korndörfer et al. 2015). There is also a marked difference in the extent people are willing to penalize those perceived as stingy (Henrich et al. 2005). Interestingly, there is a link between readiness to invest generously and the desire to punish those who do not invest (Henrich et al. 2006).

The influence of religion has also been observed. Societies where members profess Christianity or Islam give more in the first interaction in the dictator and ultimatum games—although not the commons game—than those who belong to local religious groups. (None of the ethnic groups was from a Hindu or Buddhist region.)

This connection between religion and cooperation reinforces the idea that religion has developed into a sort of support for increased cooperation in complex human societies—religions can promote prosocial behaviors. Scientists warn, however, against drawing too many conclusions from these particular results since the results on religion are not unequivocal.

Summing up the studies, it appears that 'too much' cooperation is common chiefly in modern, Western societies—the smallest scale societies play more like economists and chimpanzees. This holds even for *secular* Western states; religion diminishes in importance as the degree of democracy and education (Atkinson and Bourrat 2011) increases. If exaggerated cooperation were a fundamental human quality it would be spread across the world and about as common in every kind of society. But it isn't: cooperating 'too much' is simply not a universal human trait—norms *do* differ between societies.

Societies that live in roughly what we assume is the original lifestyle—hunter-gatherers in small groups of people who are related or know each other—demonstrate *least* cooperation and show *least* desire to punish those who do not cooperate. Mankind may be the paragon of cooperation but not all peoples possess this attribute that so perplexes economists and biologists. We are different in modern societies probably because we live in modern societies (Mellis and Semmanm 2010).

Since it is clear that cultural machinations change the propensity to cooperate, it must be possible to improve cooperation. Cooperation research thus provides hope for all the commons that humans administer.

But do these games really tell us about human nature? British games theorist Ken Binmore has a crass observation on how the dictator game relates to reality: 'The dictator game is the game we have the most data on in the world. At the moment, I have money in my wallet. I don't give them to you.'[4] Does this make the entire approach unrealistic?

Actually, no. Someone who has taken this kind of game to a higher level is the winner of the 2009 Nobel Memorial Prize in Economics, Elinor Ostrom.[5] Together with other researchers, she documented hundreds of examples where the tragedy of the commons has either occurred or been avoided. She compared these observations with the results of various game experiments, principally of commons games called *common pool resource games*. By alternating experiments with field studies, she arrived at a list of eight principles needed for commons to be properly managed (Ostrom 1990):

1. Clearly defined boundaries are necessary for the resource itself and its users;
2. Rules regarding the appropriation and provision of common resources must be adapted to local conditions;
3. Most resource users must be able to participate in the collective decision-making process;
4. Effective monitoring of the resource is needed;

[4]Personal communication.

[5]The Sveriges Riksbank Prize in Economic Sciences in Memory of Alfred Nobel 2009 was divided equally between Elinor Ostrom 'for her analysis of economic governance, especially the commons' and Oliver E. Williamson 'for his analysis of economic governance, especially the boundaries of the firm'.

5. A scale of graduated sanctions must be in place for resource users who violate community rules;
6. Quick and easily accessible mechanisms of conflict resolution are necessary;
7. Self-determination of the community shall be recognized by higher-level authorities;
8. In the case of larger common-pool resources, organization is needed in the form of multiple layers where the management of local resources takes place at the base level according to the seven previous principles.

Note that penalties and control are vital components—thus self-organization ('invisible hand') explanations are not sufficient—but also personal responsibility and community spirit. Ostrom's research shows above all that in real-life situations where problems of commons have been resolved, good institutions have been integral. An 'institution', in Ostrom's definition, is roughly the collection of rules that govern use of the resource. Locally agreed rules solve the commons problem.

The question is how to apply Ostrom's ideas on a global level. The resources of the sea and the biosphere are ultimate commons that we are all responsible for managing. The sea, like the land surface and the atmosphere, is well defined and risks destruction if not managed.

Ostrom believes, as her list manifests, that the management of large resources should be organized on several levels. What, then, should be the lowest level of organization for the sea, land and atmosphere? It is difficult to ban those who break agreed rules by (mis-) using resources. It is possible to punish environmental villains, but monitoring and risk of punishment need to be globally unified. We are far from this situation, which is tragically demonstrated by overfishing and global warming.

One problem is the atomization of humankind into so many groups. Cooperation *within* local groups seems to be inversely related to the degree of cooperation *between* groups. Investigating this, American economist Samuel Bowles constructed computer models of societies. He allowed the behavior of individuals to vary between total selfishness and total cooperation while also letting inter-group behavior vary between total intolerance and total tolerance.

Two conditions stabilized and the results were not ideal. Individuals either became very selfish at the same time as group tolerance towards other groups increased, or individuals became more cooperative within the group as the group's intolerance towards other groups increased (Bowles 2008; Bowles and Gintis 2011). It has also been shown in experiments where volunteers were administered with oxytocin ('the love hormone') that the willingness to cooperate with those perceived as friends increases, but so does the inclination to be suspicious of those outside the group (de Dreu et al. 2011).

Bowles' hypothesis is that extreme cooperation has co-evolved with the impulse to wage war. And it is indeed hard to imagine war without unselfish soldiers, willing to sacrifice themselves for others and the cause—nation, creed, God or whatever. (You would be hearing 'You first!' a lot.) According to this view, strong cooperation and strong inter-group conflict are two sides of the same coin: linked opposites. Bowles calls inter-group conflict 'the midwife of cooperation.' Indeed, humans everywhere divide the world into *us* and *them* (Culotta 2012).

If this is true it would imply that man's evolutionary history is replete with war. To investigate this, Bowles studied archaeological reviews and estimated the number of deaths caused by war by recording the proportion of skeletal remains with obvious war injuries. This method underestimates rather than overestimates the number of war dead since death can also be caused by trauma to body tissue alone. Bowles arrived at a figure of approximately 14% of the dead found in graves from 16,000 to 1000 years ago had trauma resulting from armed violence. Bowles compared this with the proportion of deaths in armed conflicts in current hunter-gatherer societies. Here, too, the figure was about 14% (Bowles 2009).

As reference, Poland lost about 16% of its population in the Second World War—more than any other country. Germany lost about 9% and the United Kingdom less than 1%.[6] Since the average fatality rate from war in our evolutionary history is comparable to the relevant figures for the worst affected regions in the Second World War, it is not obtuse to see war as an important factor in human evolution (Keeley 1996). Bowles also makes a good case that the modern democratic nation state evolved as a consequence of conflicts, as after a history of bloody wars it was the only stable bureaucratic structure left standing—'national states won wars' (Bowles 2012).

But are we not benevolent by nature, as Rousseau claimed? It is not that simple. We are obviously capable of violence and peace, cooperation and deceit. This is clearly demonstrated in 20th-century history: a period of 50 years encompassed both the Nazi regime's industrial mass murder and the secular democracies' record low prevalence of violence. Secular democracies are among the most peaceful societies ever to have existed. Human behavior is determined by an interaction of genetic, environmental and cultural circumstances and is not reducible to the simple dichotomization of whether we are good or evil by nature.

To sum up this section on man's phenomenal cooperation: there are both conceivable genetic factors and conceivable environmental explanations for why we cooperate, *as there are for all other qualities*. The environmental reasons, in turn, show that we cooperate to varying degrees depending on individuality and group association, but also on *objective* and *perceived* situations.

The explanation we seek for human 'extra' cooperation seems to be genetic and cultural, collective and individual, and also situation-dependent. Finding the source of man's exaggerated cooperation is an intricate problem, vexing many contemporary researchers.

Or does the answer lie elsewhere? For some forms of cooperation, we humans are merely a background condition; merely the substratum—the hardware.

---

[6]http://en.wikipedia.org/wiki/World_War_II_casualties.

References

Atkinson, Q. D., & Bourrat, P. (2011). Beliefs about God, the afterlife and morality support the role of supernatural policing in human cooperation. *Evolution and Human Behavior, 32*, 41–49.

Berg, N., & Gigerenzer, G. (2010). As-if behavioral economics: neoclassical economics in disguise? *Munich Personal RePEc Archive Paper* No. 26586.

Bowles, S. (2008). Conflict: Altruism's midwife. *Nature, 456*, 326–327.

Bowles, S. (2009). Did warfare among ancestral hunter-gatherers affect the evolution of human social behaviors? *Science, 324*, 1293–1298.

Bowles, S. (2011). Policies designed for self-interested citizens may undermine 'the moral sentiments': evidence from economic experiments. *Science, 320*, 1605–1609.

Bowles, S. (2012). Warriors, levelers, and the role of conflict in human social evolution. *Science, 336*, 876–879.

Bowles, S., & Gintis, H. (2011). *A cooperative species: Human reciprocity and its evolution.* Princeton: Princeton University Press.

Boyd, R., Gintis, H., Bowles, S., & Richerson, P. J. (2003). The evolution of altruistic punishment. *Proceedings of the National Academy of Sciences, 100*, 3531–3535.

Burnham, T. C., & Johnson, D. D. P. (2005). The biological and evolutionary logic of human cooperation. *Analyse & Kritik, 27*, 113–135.

Cardenas, J. C., Stranlund, J., & Willis, C. (2000). Local environmental control and institutional crowding-out. *World Development, 28*, 1719–1733.

Cesarini, D., Dawes, C. T., Johannesson, M., Lichtenstein, P., & Wallace, B. (2009). Genetic variation in preferences for giving and risk taking. *The Quarterly Journal of Economics, 124*, 809–842.

Culotta, E. (2012). Roots of racism. *Science, 336*, 825–827.

Darwin, C. (1871). *The descent of man, and selection in relation to sex.* London: John Murray.

de Dreu, C. K. W., Greer, L. L., Van Kleef, G. A., Shalvi, S., & Handgraaf, M. J. J. (2011). Oxytocin promotes human ethnocentrism. *Proceedings of the National Academy of Sciences, 108*, 1262–1266.

Dreber, A., Rand, D. G., Fudenberg, D., & Nowak, M. A. (2008). Winners don't punish. *Nature, 452*, 348–351.

Enquist, M., & Leimar, O. (1993). The evolution of cooperation in mobile organisms. *Animal Behaviour, 45*, 747–757.

Fehr, E. (2004). Don't lose your reputation. *Nature, 432*, 449–450.

Fehr, E., Fischbacher, U., & Gächter, S. (2002). Strong reciprocity, human cooperation and the enforcement of social norms. *Human Nature, 13*, 1–25.

Fehr, E., & Gächter, S. (2002). Altruistic punishment in humans. *Nature, 415*, 137–140.

Frank, R., Gilovich, T., & Redan, D. (1993). Does studying economics inhibit cooperation? *Journal of Economic Perspectives, 7*, 159–171.

Friedman, M. (1953). *Essays in positive economics.* Chicago: University of Chicago Press.

Gächter, S., Herrmann, B., & Thöni, C. (2010). Culture and cooperation. *Philosophical Transactions of the Royal Society B: Biological Sciences, 365*, 2651–2661.

Gneezy, U., & Rustichini, A. (2000). A fine is a price. *Journal of Legal Studies, 29*, 1–18.

Hardin, G. (1968). The tragedy of the commons. *Science, 162*, 1243–1248.

Hauser, M. (2006). *Moral minds: How nature designed our universal sense of right and wrong.* New York: HarperCollins.

Henrich, J. (2016). *The secret of our success: How culture is driving human evolution, domesticating our species, and making us smarter.* Princeton: Princeton University Press.

Henrich, J., Boyd, R., Bowles, S., Camerer, C., Fehr, E., Gintis, H., et al. (2005). 'Economic man' in cross-cultural perspective: Behavioral experiments in 15 small-scale societies. *Behavioral and Brain Sciences, 28*, 795–855.

Henrich, N., & Henrich, J. (2007). *Why humans cooperate.* Oxford: Oxford University Press.

Henrich, J., McElreath, R., Barr, A., Ensminger, J., Barrett, C., Bolyanatz, A., et al. (2006). Costly punishment across human societies. *Science, 312*, 1767–1770.
Herrmann, B., Thöni, C., & Gächter, S. (2008). Antisocial punishment across societies. *Science, 319*, 1362–1367.
Hoffman, E., McCabe, K., Shachat, K., & Smith, V. (1994). Preferences, property rights, and anonymity in bargaining games. *Games and Economic Behavior, 7*, 346–380.
Jensen, K., Call, J., & Tomasello, M. (2007). Chimpanzees are rational maximizers in an ultimatum game. *Science, 318*, 107–109.
Kahneman, D. (2011). *Thinking, Fast and Slow*. New York: Farrar, Straus and Giroux.
Kahneman, D., & Tversky, A. (1979). Prospect theory: An analysis of decision under risk. *Econometrica XLVII*, 263–291.
Keeley, L. H. (1996). *War before civilization*. Oxford: Oxford University Press.
Kitcher, P. (2011). *The ethical project*. Cambridge: Harvard University Press.
Korndörfer, M., Egloff, B., & Schmukle, S. C. (2015). A large scale test of the effect of social class on prosocial behavior. *PLoS ONE, 10*, e0133193.
Laland, K. N., & Brown, G. R. (2002). *Sense & nonsense*. Oxford: Oxford University Press.
Liberman, V., Samuels, S. M., & Ross, L. (2004). The name of the game: Predictive power of reputations versus situational labels in determining Prisoners' Dilemma game moves. *Personality and Social Psychology Bulletin, 30*, 1175–1185.
Lindenfors, P., Nunn, C. L., & Barton, R. A. (2007). Primate brain architecture and selection in relation to sex. *BMC Biology, 5*, 20.
Mellis, A. P., & Semmanm, D. (2010). How is human cooperation different? *Philosophical Transactions of the Royal Society, Series B: Biological Sciences, 365*, 2663–2674.
Nee, N., & Bailenson, J. (2007). The Protheus effect: the effect of transformed self-representation on behavior. *Human Communication Research, 33*, 271–290.
Nikiforakis, N. (2008). Punishment and counter-punishment in public good games: can we really govern ourselves? *Journal of Public Economics, 92*, 91–112.
Nowak, M. A., & Highfield, R. (2011). *Supercooperators*. London: Canon Gate.
Oda, R., Kato, Y., & Hiraishi, K. (2015). The watching-eye effect on prosocial lying. *Evolutionary Psychology, 13*, 1–5.
Ostrom, E. (1990). *Governing the commons: The evolution of institutions for collective action*. Cambridge: Cambridge University Press.
Ostrom, E. (1998). A behavioral approach to the rational choice theory of collective action. *The American Political Science Review, 92*, 1–22.
Ostrom, E., Walker, J., & Garnder, R. (1992). Covenants with and without a sword: Self-governance is possible. *American Political Science Review, 86*, 404–417.
Panchanathan, K., & Boyd, R. (2004). Indirect reciprocity can stabilize cooperation without the second-order freeloader problem. *Nature, 432*, 499–502.
Rand, D. G., Greene, J. D., & Nowak, M. A. (2012). Spontaneous giving and calculated greed. *Nature, 489*, 427–430.
Rankin, D. J. (2011). The social side of Homo economicus. *Trends in Ecology and Evolution, 26*, 1–3.
Rankin, D. J., Bargum, K., & Kokko, H. (2007). The tragedy of the commons in evolutionary biology. *Trends in Ecology and Evolution, 22*, 643–651.
Read, L. (1958). *I, Pencil: My family tree*. Irvington-on-Hudson: Foundation for Economic Education.
Ricardo, D. (1817). *On the principles of political economy and taxation*. http://www.econlib.org/library/Ricardo/ricP.html
Ridley, M. (1996). *The origins of virtue*. London: Penguin.
Ridley, M. (2010). *The rational optimist*. London: Fourth Estate Ltd.
Shakespeare, W. (1600, approximately). Hamlet.
Shermer, M. (2008). *The mind of the market*. New York: Henry Holt and Company.
Simon, H. A. (1957). *Models of man*. Hoboken: John Wiley.

Smith, A. (1904). *An inquiry into the nature and causes of the wealth of nations*. London: Methuen and Co., Ltd.

Tomasello, M. (2009). *Why we cooperate*. Boston: Boston Review.

Warneken, F. (2013). Altruistic behaviors from a developmental and comparative perspective. In: K. Sterelny, R. Joyce, B. Calcott, & B. Fraser (Ed.), *Cooperation and its evolution* (pp. 399–423). Cambridge: MIT Press.

Yamagishi, T. (1986). The provision of a sanctioning system as a public good. *Journal of Personality and Social Psychology, 51*, 110–116.

Zahavi, A., & Zahavi, A. (1997). *The handicap principle: A missing piece of darwin's puzzle*. Oxford: Oxford University Press.

Language

Abstract Language is a uniquely human endeavor that both *is* cultural transmission and *enables* cultural transmission. The traits in language that animals cannot handle are presented: symbols composed of symbols and grammar (i.e. semantics and syntax). I also briefly discuss the evolution of language and a theory of my own on language evolution. It is pointed out that language enables a novel replicator: ideas, thoughts, or "memes". I discuss why "meme" may not be such a good word (it sounds like we know what we are talking about when we don't).

> There is no mode of action, no form of emotion, that we do not share with the lower animals. It is only by language that we rise above them, or above each other - by language, which is the parent, and not the child, of thought.
>
> Wilde: The critic as artist (1891).

There is a fundamental difference between cooperation in humans and cooperation in other animals. We humans can transmit information between ourselves—we can gossip and gang together, give someone a scolding and bellow instructions, we can give compliments and flattering praise. Through language we can agree on complex ethical rules; we can transmit traditions and laws and factual information.

With the aid of language you can easily solve the dilemmas that arose in the games discussed in the last chapter. Permitting discussion before playing allows people to almost *always* solve the dilemmas (Bower 1990; Ostrom et al. 1992; Dawes et al. 1990)—allowing gossip is more effective in maintaining cooperation than is punishment (Wu et al. 2016). The only reason the dilemmas are dilemmas is that information transfer is artificially obstructed. No other animal has this possibility of information sharing—no other animal has language.

For communication *without* language you need a signaling individual and a recipient who can react to the signal. For a scientist to know that communication has occurred, it is also necessary that the recipient changes its behavior. This kind of communication, when it works, can thus be described as a form of manipulation; the signaler manipulates the recipient (Dawkins and Krebs 1978).

A simple case of this is bacteria who can communicate using chemical substances. Some bacteria will release signal substances in the environment as a reaction

P. Lindenfors, *For Whose Benefit?*, DOI 10.1007/978-3-319-50874-0_8

to environmental cues. If the bacterium is subject to stress because of starvation it can release a substance, a signal which is ignored by the surrounding bacteria (i.e. they do not change behavior) until the substance reaches a certain threshold concentration. When this threshold is reached, other bacteria will turn on or off certain genes and also release the substance that further increases the concentration. The signaling has the result that a number of bacteria disperse (or aggregate) and by so doing, more effectively colonize and exploit resources in the environment.

Bacterial communication almost always occurs between closely related cells and can therefore be explained using Hamilton's rule for kin selection. There is a common genetic interest in cooperating. Also, there is often direct mutual benefit. Both the bacteria that signal and those that are manipulated benefit from this communication; the manipulated bacteria benefit from being manipulated (otherwise there would be selection against reacting to the chemical).

But more complex cases of communication outside language also exist. A male bird who sings—it is almost exclusively males who sing—transmits information to neighboring male birds; that the territory is taken and that they therefore should stay away. But he also communicates his location to the females and that they are welcome to come and stay with him, should they so wish.

Many birds have several calls. There are special contact calls and warning calls, and hungry young can have special begging calls. There is also a phenomenon called subsongs: young birds practicing their songs softly, softly. In late winter, blackbirds can be spotted hidden inside a tree, practicing quietly for the mating season (Fitch 2010).

There is a similar variation of vocalizations in vervet monkeys—a species of monkey from Southern Africa—where different alarm calls are used for different species of predators (Cheney and Seyfarth 1990). This association between a certain sound and a certain phenomenon (*Danger on the ground! Danger in the air!*) is viewed as surprisingly advanced for an animal. Interestingly, a bird, the yellow-casqued hornbill, has learnt to interpret the warning calls of vervet monkeys and hides when it hears the call for *Danger in the air!* but not when it hears the call for *Danger on the ground!* Eagles are a definite threat to hornbills, but leopards aren't (Rainey et al. 2003).

But communication does not have to occur through sound. For example, one of the most advanced examples of information transfer besides human language occurs in bees who tell their fellow bees in the hive about the direction to a food source using an intricate 'waggle dance' (von Frisch 1967). When a bee has located a resource, it informs other bees in the hive of this through running an elongated figure 8-loop, 'waggling' during the central path through the 8. The direction and duration of the 'waggling' indicate direction and distance to the resource. Each bee seems to weigh this socially transmitted information to individually gathered information, and in the end individual bees utilize information from the waggle dance only about 10% of the time (Dornhaus and Chittka 2014), more often when resources are scarce (Grüter et al. 2008).

However, all examples of information transfer in the animal world, using gestures, sounds, chemicals, facial expressions, singing or dances, are not even close to the complexity of human language. What is it about human language that is so complex? And why do we have language ability while other animals don't?

The question has been raised if the differences in language ability between humans and other animals are due to the fact that humans grow up immersed in language. Everyone speaks: parents, teachers and friends; even the grumpy old men who scolded you as a child used language to get the message across. Certainly, animals who grow up immersed in language can also learn a great deal of its meanings. Dogs can learn to associate specific words with specific items or activities. The reigning champion is Chaser the Border collie who can distinguish between words for 1022 different toys (Pilley and Reid 2011).

So what would happen if you took a very intelligent animal, for example a chimpanzee, and tried to teach it language from birth? Would it be able learn language in full, or at least in part? Scientists tried for a long time to teach chimpanzees to communicate with language, but only by the late 1960s could it be reported that a chimpanzee female by the name of Washoe had cracked the code and learnt to communicate using several hundred words. The researchers who had managed the feat of teaching Washoe were Allen and Beatrice Gardner, who had trained Washoe in a trailer in their backyard.

The key to success for the Gardners was that they didn't try to teach Washoe spoken language, but sign language. It is impossible to teach chimpanzees to speak for purely physiological reasons since they lack the vocal apparatus that humans possess, including the extreme control of breathing that is necessary.

Washoe didn't react only to commands but could also, using sign language, put together brief sentences like 'give candy' or 'listen dog'. A famous story relates when Washoe was out on a trip and saw a swan for the first time. Excitedly she used two words that she never had connected before and signed 'water bird'. (However, some say that as she was just using two words she already knew, it shouldn't count as a new word).

Washoe had many successors, among others a gorilla named Koko, an African grey parrot named Alex, and a chimp named Nim Chimpsky—named after the great linguist Noam Chomsky who claims that language is a uniquely human trait. The researcher training Chimpsky, the psychologist Herbert Terrace, wanted to systematically prove that Noam Chomsky was wrong and that this difference between man and animal didn't exist.

Terrace and his colleagues trained Chimpsky in the way Washoe had been trained, and achieved the same success. To investigate if Chimpsky could understand grammar the researchers conducted several careful analyzes of video tapes of Chimpsky communicating. This was an important step, because just putting two words together can mean completely different things.

'Grapes me' can, for example, mean 'give me grapes', but equally well 'I'll get the grapes'. To confirm that Chimpsky really understood language, Terrace had to show that Chimpsky could understand the difference; that Chipmsky knew grammar.

But after long hours and careful analyses Terrace had to conclude that Chimpsky didn't understand grammar—the opposite result to the one that he was expecting.

Chimpsky had reacted to subtle signs from the trainers and learnt to use language through endless repetition, not through understanding.

Researchers in the field were understandably upset with Terrace. But it turned out that he most probably was right. Careful analysis of the chimpanzees' use of language showed no grammatical comprehension.

Yet one project continued to show positive results. A bonobo named Kanzi was taught to use a keyboard with 256 symbols. Kanzi quickly learnt to use them to put together simple sentences. His abilities were tested by providing commands through the use of the symbols while ensuring that Kanzi couldn't actually see any human being. This to make sure that no other, subtle signs were communicated that enabled Kanzi to make short-cuts to understand meaning. And commands were put together using already known symbols, but in internal relationships that Kanzi hadn't encountered before to make sure he wouldn't just memorize sequences.

Kanzi passed these tests impressively. 660 different sequences were tested and Kanzi almost always did as instructed. Again the scientific community was impressed—this was proof that at least bonobos could master language, right? But a psychologist by named Clive Wynne has analyzed these results more closely and shown that even this case didn't involve real grammatical comprehension.

It turns out that only 21 of the 600 commands actually tested grammatical understanding. If you give the command 'Put the sponge in the bowl' it is easily deduced from the words 'sponge' and 'bowl' what is to be put where, you simply cannot put a bowl in a sponge. But if you give the command 'pour the juice in the milk' it becomes more difficult—you need grammatical comprehension to get it right. In the 21 cases that actually tested grammar, Kanzi responded correctly only 57% of the trials—almost random.

There was also another problem. The scientists who tested Kanzi approved some dubious cases. When instructed to 'Let the dog bite the snake' and 'Let the snake bite the dog' Kanzi in both cases put the toy snake in the mouth of the toy dog, but was scored 'correct' both times (Wynne 2007). No ape has been able to consistently pass tests like these, nor been able to combine words to phrases or express temporal propositions using language.

This doesn't show that apes cannot associate symbols with items, but it shows that they seem to lack the ability to learn grammar—syntax. Thus, we can study language properly only in humans. Rudimentary language in animals is at such a level of simplicity that it helps our understanding of language very little to study animal communication. This means that since the invention of complex language, we have been talking to ourselves.

The Structure of Human Language

What makes human language so fiendishly difficult that no animal can learn it but simultaneously so simple that any child can? There seems to be two aspects to this: an ability to construct what can be termed 'meta-symbols'—symbols that only relate to other symbols, and an ability to use and understand grammar, to change the meaning of a sentence by re-ordering its components. As we will see, however, these may be two aspects of the same problem.

One of the building blocks of language is the arbitrary connection between a word and the object that the word is referring to. For example, a butterfly is famously called 'papillion' in French, 'Schmetterling' in German, 'butterfly' in English and 'fjäril' in Swedish—all arbitrarily connecting a number of interconnected sounds with a group of animals. The scientific understanding of meaning, the study of content and interpretation of these types of symbol systems, is called semantics.

As we just saw, many animals can associate words with objects and actions, and can learn many such associations. But language quickly becomes more complex. How, for example, should one know that 'butterfly' applies to some fluttering animals with colored wings but not others? That the group 'butterfly' belongs to the larger group 'insects' in the same way that 'monkeys' belongs to 'mammals'. These types of generalizations and sets within sets occur on many levels in language; language contains a hierarchy of meanings.

Terrence Deacon, an American anthropologist, has pointed out that one really hard problem for animals is to handle words made out of other words. Adhering to the terminology used in his book 'The Symbolic Species', animals can handle icons (that refer to objects or acts by way of resemblance; as pictures resembles objects) and indices (that refer to objects or acts through direct connection; smoke is an index for fire), but not symbols (words that can be used without such context and that can only be described using other words, like the word 'whatever') (Deacon 1997).

We humans have dictionaries where we use words to explain other words—with no other aid than words. Take the word 'apropos'. According to the Merriam-Webster dictionary it means '1: at an opportune time, or 2: by way of interjection or further comment: with regard to the present topic'. How would you even begin to explain this to a chimpanzee? No animal has passed this 'symbolic threshold' but must relate every new word they are to learn with a concrete object or a concrete activity. Animals cannot fully master semantics.

The other building block of language is, as we saw before, syntax, or grammar. This describes how to associate elements to working sentences. It is in the use of syntax that humans already at the ripe age of two surpass even the most advanced sign-language-using apes. In syntax, the ordering of words is central. But there is no unanimity in the world's languages as to exactly *which* order; there are a lot of systems.

In the 1950s, linguist Noam Chomsky introduced the concept of 'generative grammar' to systematically describe the grammar in different languages (Chomsky 1957). The grammatical rules for how words are combined to construct meaningful sentences are one of the things that makes language so useful. It doesn't really matter *what* rules apply, as long as they are shared between language users. Chomsky described how grammar from all languages can be described and understood within a single framework. He further suggested that the human ability to learn grammar is hard-wired into our brains—*universal grammar*. Is this true, and if so, where did this ability come from?

Recently, a significant advance has been made concerning the difference between humans and other animals that may shed light to what ability that is missing for animals to master grammar. While it long has been realized that many abilities set humans apart from other animals, such as language, culture and an ability for problem solving, it has proven very difficult to identify any one, specific cognitive trait that humans have that animals don't—though there have been a lot of failed claims. In a recent project, a novel such ability has been identified: other animals have a hard time to discriminate ordered sequences of stimuli if the stimuli are spaced too many seconds apart.

Three researchers from Stockholm University are behind this finding, psychologist Stefano Ghirlanda and biologists Johan Lind and Magnus Enquist (Girlanda et al. *in press* 2017). Their results do not negate the fact that animals are capable of discriminating between *specific* sequences that they have evolved to discriminate between; the find is that animals, except humans, are not able to discriminate between *arbitrary* sequences.

The three researchers investigated the hypothesis that the lack of the ability to discriminate between arbitrary stimulus sequences has to do with fading memory traces. In 108 experiments reviewed, the difficulty for animals lies in discriminating simple sequences such as these: AAA, AAB, ABA, ABB, BAA, BAB, BBA and BBB, if the spacing between the signals in the sequences are too long.

The researchers suggest that this is because the memory of earlier events fade over time too quickly to remember early stimuli, and thus the animals discriminate according to only the last signal. Animals were, for example, found to discriminate more easily between AAB and AAA than between AAB and BAB—the success rate for the latter type of sequences was extremely low even after long training. The longer time that elapsed between the stimuli, the worse animals did. The results held for a large number of species.

Humans, on the other hand, do not have this problem. Sequences of stimuli seem to be stored some other way in human brains, where temporal sequence seems to be part of the information stored. Another way to state the same observation is that animals do not seem to have an internal representation of time. While this ability to distinguish stimuli sequences potentially can explain many differences between humans and other animals, it may also provide answers to several of the puzzles surrounding language.

First, it indicates why humans, but not other animals, can master grammar. Grammar, broadly speaking, transmits information using order. The meaning of sentences change when the order of morphemes (the smallest units carrying information) change. It doesn't matter exactly *what* the order is—grammar is different in different languages—but it matters a lot *that* there is an order which carries information.

Second, it potentially also explains why humans, but no other animals, can construct words out of other words—how we crossed Terrence Deacon's 'symbolic threshold'. This ability becomes possible with general sequence processing capabilities described above, because an ability to discriminate between, for example, AAB and BAB means that stimuli ('A' and 'B') are grouped into larger structures ('AAB' and 'BAB') that take on different meanings, hence making it possible to construct symbols from icons and indices—and to construct symbols from other symbols.

Third, the same capacity to group stimuli potentially explains why humans, but not other animals, can embed clauses recursively. According to Noam Chomsky, this ability depends on a capacity to perform a 'merge' operation—to take any two (or more) syntactic elements and combine them into a new expression; an expression that then in itself can be used as a single syntactic element (Berwick and Chomsky 2016).

This concept may be hard to comprehend. Berkeley linguist Marc Ettlinger has helpfully shared the following example on Quora to make it more understandable, where a sentence is 'merged' and inserted as a single syntactic element in another sentence.

The cat meowed.
The cat [the dog chased] meowed.
The cat [the dog [the man hit] chased] meowed.

This is what is termed *recursion* in linguistics—it's not just looping or repetition, but 'the embedding of a structure (sentence) within another structure of the same type' (Ettlinger 2014).

A potential first step on the way to full language capacity could thus be the evolution of this seemingly very basic better ability to discriminate ordered sequences of general stimuli. Note also that this is a trait that can evolve and improve gradually.

With a limited number of symbols and a common syntax sentences can be constructed describing anything—with limited means you can express an unlimited number of meanings. This is why it was so important to check if the apes in the language experiments had an understanding of grammar. Without grammar there is no dynamic language, only static associations between words and objects, traits or actions.

What becomes possible with language, something that animals cannot do and something that is not possible with other forms of communication (for example making faces or through music), is to make factual statements with a truth value; statements that can be true or false. With language you can state propositions about reality to other individuals (Fitch 2010) and thus share factual information.

Animals cannot overcome 'the symbolic threshold' and cannot learn to grasp the ordering issues involved in learning grammar; animals are limited in their abilities both concerning semantics and syntax. How come humans have these abilities?

The Evolution of Language

Chomsky described how grammar has the same rough structure in all languages ever studied—they can all be described using the same grammatical terms. Also, we all use approximately the same number of words in our daily lives. There are substantial problems with counting words in languages, but random samples from lists of words in well-documented languages seem to show that people have a passive vocabulary of about 80,000 words. However, of these, we use only 4000–8000 words daily. Since only about 200 of the world's 6000 languages are written, there is admittedly a limited number of languages that can be studied in this way (Parkvall 2009). And there is contradictory research indicating differences in the size of vocabularies (Henrich 2016).

Nevertheless, how come the number of words utilized hasn't just kept increasing during cultural evolution? Why doesn't the number of words commonly used differ immensely between modern societies and hunter-gatherer societies? Don't we live in much more complex circumstances now?

Presumably, this has something to do with a limitation in our brains when it comes to keeping track of more than this aggregate of words. But even more probable is that there is a practical limit, in that a larger number of words does not necessarily increase communicative potential. For example, Shakespeare used about 18,000 words in his plays, and he used almost the same words in Hamlet as in Othello. Despite this similarity the two plays are completely different (Ridley 2009).

One possible conclusion from this is that languages on some basic level have finished evolving and are optimized for humans—languages can't get any better than they are. Words and grammar can change, for sure—this happens all the time —but the number of commonly used words doesn't grow and the basic grammatical structures are stable over time (which is why we can describe all languages using the same grammatical terms). The question is, how did we get to this point?

First, humans had no language—now we do. For a long time, this was the full extent of our knowledge of language evolution. Such a lack of basic facts easily leads to speculative and untestable theories. In 1866 this situation, with many wild theories floating around, led the *Société de Linguistique de Paris*—the Linguistic Society of Paris—to famously ban speculation about language origins pending more information (though this seems not to be a fully correct description of what really happened).

We now know more about almost everything than what was known in 1866. But has our knowledge of language evolution progressed in a manner that means that we can disregard the Paris Society's ban? Judging from the flood of articles and books on language evolution, many researchers think so. Most researchers (with the possible exception of some scientists who still are trying to teach language to apes) now agree that grammatically advanced language is a uniquely human skill. But besides that, scientific views on language evolution are far from unanimous.

Theories of language evolution are numerous. Language may have evolved as a form of verbal grooming when human groups became too large to allow actual communal grooming. Others have suggested that language evolved together with tool use, or that language was necessary for the social contract between males and females on division of labor (men hunt and females gather in all contemporary hunter-gatherer societies).

Did language evolve as a status badge where those who could transmit information were accorded high status? Or as a by-product of singing, as a trait favoring group cohesion, to comfort babies when these were put down to gather, or as an adaptation to hunting? Finally, some have suggested that language primarily emerged as a form of costly ornament that men use to impress females. (But surely, not only men are able to talk?) Or to be used as a thinking aid where communication was only incidental—an emergent effect? (Számadó and Szathmáry 2006).

All these theories have their weaknesses, especially the last one. Why would there be a need to translate inner thoughts to a representation outside the brain when thinking occurs within the brain anyway? No, if one thing is certain about language evolution, it is that language emerged to facilitate communication—this is what language is all about. As the Swedish cognition researcher Peter Gärdenfors has put it: 'Dialogue must come before monologue' (Gärdenfors 2000).

In some sense, it does not really matter for our discussion on human cooperation *why* you need to communicate, or *when* language evolved. The important thing is to realize *that* it happened. With a dynamic language, once you can communicate something, you can communicate many things. But not knowing everything is not the same as not knowing anything, there is some knowledge that researchers agree upon.

All peoples can communicate using language; all peoples have the same language centers in the brain, the same extreme control of breathing, the same precise command of the tongue and lips, and the same arrangement of the language apparatus in the throat. Language ability thus certainly has a common origin for all humans. But when did it emerge?

To a certain extent it is possible to look at the fossil record of when the morphological changes that are needed for spoken language emerged to at least establish an approximate date. It seems that Neanderthals had a hyoid bone practically identical to ours (which indicates a descended larynx) (Arensburg et al. 1990) and an enlarged hypoglossal canal—the canal that leads nerves from the brain to the chest (indicating precise control of breathing) (Kay et al. 1998) and at least one crucial gene that is necessary for language (Krause et al. 2007). Thus, Neanderthals could probably talk.

Previously, a descended larynx was regarded as crucial for human speech and it was marveled at the increased risk of choking that this entailed. Research now indicates that the larynx is movable in several animals—the larynx descends when animals call—so the ability to produce calls doesn't seem limited by the larynx. There is also no data on choking in animals—and humans only rarely choke—so we simply do not know if there really is a substantial cost of a descended larynx for humans (Fitch 2010). There are also other animals that have specialized skin folds for sound production—vocal cords—so that is no constraint either. Thus, it is a bit tricky to date language in the fossil record, but the basic ability may be as old as the genus *Homo*—a couple of million years.

Humanity has had such benefit from being able to transmit information that it is important to stress that there also is a cost. There are many adaptations that are needed for spoken language to work. Breath needs to be precisely controlled, the tongue and the lips must be controlled with incredible precision and speed, and the brain must be more powerful than ever to handle this advanced form of motor control. In humans, a third of the motor control-center in the cortex is used to govern the tongue, mouth, face and thread. In apes this proportion is just a tenth (Gärdenfors 2000).

Add to this, specialized brain sections for handling language production and language comprehension: Broca's and Wernicke's areas. Coding and de-coding sound differences that occur as rapidly as they do in language demands a lot from an information processing unit.

Note, however, that these brain parts need not have evolved as language modules, but may equally likely be parts of the brain that are high-jacked by language during development. There is an identified part of the brain that is used for reading in almost all humans, and we *know* we didn't evolve reading capacity. Just because something is biological doesn't mean it is genetic—what evolved was probably a brain with a *capacity* to become high-jacked.

A large brain needs a lot of energy. In other apes, it is estimated that about 5% of the body's energy is being used in the brain. In humans, this proportion is 20%. No one believes that all this enlargement is due to language, but the fact that two rather larger parts of the brain become dedicated to language, and that even more parts are involved in speech production and language comprehension, indicates the importance of language for brain evolution in general. We shouldn't forget increased maternal mortality due to the larger head size of infants. Language has come with substantial costs.

The advantage, on the other hand, is crystal clear. With language, all forms of human cooperation will be easier, since you can express your thoughts and transmit them in a mutually agreed manner to other individuals (Smith 2010). Genes for language capacities and cultural evolution of language thus likely co-evolved. But how did humans cross the threshold where advantages outweighed costs?

The Green Beards of Language

As we have already seen, cooperation demands special circumstances to evolve. This is also true for circumstances that improve cooperation. Language places extensive demands on adaptations both in speech and comprehension. To mumble wishes into thin air may work in contact with supernatural beings, but as a form of cooperation between humans it is insufficient—someone needs to be listening.

This means that we have to explain a trait—language—that completely builds upon cooperation and where capacity to produce a signal (speech) and a capacity to interpret that signal (comprehension) must evolve together. Earlier chapters explained the biological evolution of cooperation with the help of kin selection and reciprocity. Are these theories practical for understanding the evolution of language as well?

I have published my own theory on language evolution in an attempt to clarify this (but this is thus far only a hypothesis, so read the next section critically, please) (Lindenfors 2013). Remember the Green Beards? This was Hamilton's insight that if an individual could recognize the trait 'cooperation' in another individual through the use of some sort of signal that was impossible to fake—for example a Green Beard—then the evolution of cooperation would be much easier because the willingness to take on a cost would be aimed towards cooperative individuals at all times, and never at cheaters.

Now think about language. How do you know that someone else understands English?

If you speak English and the other person understands and answers, you have quickly assessed each other's knowledge of language. For the trait 'English' the signal is completely honest and impossible to fake. You cannot *pretend* to know how to speak English (if you have learnt it you know it, so you can't be faking even if you speak with an accent). You don't get very far with *pretending* to understand either, sooner or later you will be exposed, especially in two-person interactions. These encoding and decoding systems had to evolve simultaneously, through mutual benefits.

If you look at it more closely, language transfers information on at least three levels; (1) what is said, (2) how it is said (what language is used), and (3) that it is said (that speaker and listener both possess the ability to use language). The use of language is a form of honest cooperation on two of these levels; not necessarily on *what* is said, which can be deceitful, but always on *how* it is said and *that* it is said.

Crucially, both the biological (point 1) and cultural aspects (point 2, but not point 3) of language are equipped with Green Beard recognition systems. The biological capacities required for language signal their presence through speech and understanding. This is signaling that cannot be invaded by 'false' Green Beards because the traits and the signal of their presence are one and the same. You can't signal being able to talk without being able to talk.

However, the real usefulness of language comes from its potential to convey an infinite number of meanings through the dynamic handling of symbols—through language itself. But any specific language also signals its presence to others through usage and understanding. You cannot signal being able to speak and understand English to an English speaker without being able to speak and understand English. Thus, languages themselves cannot be invaded by false Green Beards because, again, the trait and the signal of its presence are one and the same.

These 'twin' Green Beards associated with the biological and cultural abilities seem to be unique for language. Concerning biological traits, only a small number of examples of traits equipped with the Green Beard recognition have been identified (Gardner and West 2010). There may also be other cultural traits that have Green Beards (investigation is lacking) but both biologically and culturally simultaneously? In that regard, language is unique.

The problem with Green Beards for other cultural traits is that they can be exchanged too easily. If you choose to cooperate with someone carrying a certain marker—e.g. people with necklaces with crossed bones that shake feet instead of shaking hands—then these Green Beards will work in the short term. But you can easily cheat on such signals by refraining from reciprocation or changing traits, the latter either by taking off the necklace or donning a fake one. Language is different, because most people retain their language their whole life.

On top of that, languages have the unique property of not making cooperation a condition for the trait to be benefited. When arguing, for example, you use (hopefully, almost only) language to get your point across. If you tell someone to 'F-k off!' you are cooperating in using language and which language to use—even if you would rather terminate cooperation with the other individual! You can cheat on *what* is said, but never on *how* it is said or *that* it is said. That is my hypothesis on language evolution (Lindenfors 2013). As an extension, I would like to add that I believe the ability to utter falsehoods evolved *after* language abilities—but that is *really* speculative.

The Second Replicator

Time to return to Richard Dawkins' metaphor about the selfish gene. It was formulated to ease the understanding of how genetically determined cooperation could evolve. It is easier to understand how any individual can sacrifice part of their own reproduction if you see genes as selfish, with the sole interest of getting copied.

Now use the same metaphorical thinking on language. Instead of genes, view language elements as being selfish. A 'selfish language element': a language element that 'acts' as if it was selfish favors its own existence (survival) and propagation (reproduction). Whenever people reach an agreement, fight, bargain, plea, threaten or do anything that requires verbal communication, language is used to transmit information between the involved parties.

This rapidly becomes a self-amplifying process. Once started, communication via language is difficult to stop. Even when you yell at someone to 'Shut up!' you do so using language. So much of human social life is dependent on verbal communication that there is only one way for selection to go: towards even more expressive language.

The American anthropologist and anthropologist Terrence Deacon and others have suggested that this is a powerful enough process to explain the expansion of the human brain and the evolution of all components of the language apparatus (Deacon 1997). Language can have evolved together with language ability and be what has driven the evolution of the large human brain and our unique cognitive skills. Other researchers mean that it is more probable that the human brain has evolved as an answer to a whole complex of social skills they term 'cultural intelligence' (Herrmann et al. 2007).

To be able to express anything it must be possible to shuffle symbols and put them in different contexts, and also to quickly come up with new symbols and new arrangements of these to express new thoughts. Thus, language demands the ability of *free thinking*.

Humans have to have a sense of how order can change meanings (Gärdenfors 2000) and an ability to rapidly decode such meanings. A brain is needed that can handle huge amounts of information in a generalized manner and isn't locked into fixed interpretations. If the ability to manipulate symbols emerges, then simultaneously a strong evolutionary push for a capable and above all *flexible* brain emerges, to handle the shuffling of symbols and their connotations.

At the other end, pressure emerges to make language flexible enough to allow the expression of every conceivable meaning while simultaneously being simple enough to be learnt by very small children. What is more, to work, this learning has to be social. You cannot be born with a language—this would not function as a dynamic system of communication. This specification means that languages appear the way they do because they fulfill these two specifications of being practical to express every conceivable meaning and at the same time simple enough for children to learn.

We should thus expect language to work as a form of cultural amoeba that fits perfectly into the brain's convolutions: as a 'parasite' that rebuilds its host in order to fit itself. Our language has the structure it has because human brains learn language having exactly this structure. And here is where we find out why small children can learn language while chimpanzees have no reasonable chance to even come close: human language and human brains have evolved together for a very long time to fit each other.

There exists an evolutionary process termed the Baldwin effect, where a behavior that initially had to be learnt gets 'written into' the genes. The thinking here is that it is beneficial for a crucial behavior/ability to be hard-wired into the genes instead of having to be learnt—so it can't be lost to a population. Something similar has been suggested as an explanation of how language and language ability coevolved (Deacon 1997).

If so, it is a generalized ability to handle symbols that our genes carry instructions for, not a specific encoding/decoding mechanism. Children are not born with a language, but with the ability to learn a language. Chomsky may not have been so wrong on this after all, when he hypothesized that humans are born with an ability to learn grammar (Chomsky 1957). But it is not through a specific 'grammar module', but because languages have become structured the way they are because this is what fits our brains and how they learn.

Language provides us with a control panel to the brain. Using language we can manipulate both others and ourselves. We can store information and processes in our brains—we can learn and by doing so rebuild our brains. This is most probably the only feasible way to arrive at general intelligence: the possibility of being able to reprogram your own information-handling unit (Hofstadter 1979). This reprogramming makes cultural differences biological (differences in the brain) but not genetic (they are learnt, not genetically inherited).

So can language be the product of a general system of information handling that has evolved for some other reason than language? Of course it can. There are, however, signs that this is not the case, but that language is a unique adaptation in humans, an adaptation that demands unique abilities.

For example, it is possible to have a stroke and lose almost all speech ability but with intellectual abilities only slightly damaged. Often the ability that is harmed is the ability to use and understand grammar. There are also examples of the opposite, that humans with an intellectual disability can speak grammatically correct nonetheless (Deacon 1997). To some degree, intellect and language ability are separated.

But completely? No, the handling of language does have an intellectual aspect. And with a brain that can freely juggle symbols and decode complicated sequences we simultaneously get a tool that can also be used to learn other things. Or, put differently, we have the prerequisites for other traits—learnt traits—to colonize our brains and communicate with each other, outside of our control.

This is the first time since discussing genes that we have a new system in place where information can be transmitted as instructions instead of as a finished product (Blackmore 1999). This means that the transmission mechanism (language) can be protected from changes in what the information is about (ideas) in the same way as the biological transmission mechanism (genes) is protected from what happens to the biological traits. Also, language makes idealism possible—it is now possible to talk about, describe and define ideal states (triangles, the perfect growing season, quality) that do not exist other than as thoughts.

Transmitting instructions is both easier and more stable than to transmit the finished product; less information gets lost on the way. Give someone a recipe for a cake and the manufacturing process is quickly understandable, but just give the cake and the manufacturing process is nearly impossible to reconstruct. This is why you ask a good cook for the recipe and not for a copy of the dish to take home.

The stability of instruction transfer is amply illustrated by cases where we are left only with the finished product. The pyramids, the statues of Easter Island and the Nazca lines (the gigantic figures in the desert sands of Peru) are all examples of enigmatic cultural products. The instructions of how to make them got lost in the mists of history and it took a long time and hard thinking to decode how they were once made. Besides language, humans have invented many other systems for instruction transfer, for example writing, musical notation, cooking recipes and architectural drawings, all depending on our understanding of semantics and syntax.

With language a new replicator has entered the scene. Thus far all cooperation we have looked at has worked for the benefit of the initial replicator—the genes. Now there is a new component that also can be copied better or worse—the replicator Richard Dawkins has termed memes: ideas and thoughts, or basically, that which is transferred when we learn something (Dawkins 1976).

This means that language can be a possible explanation of the exaggerated form of cooperation we left more or less unexplained the last chapter. It isn't only that language helps people to cooperate better because they can express intentions, provide instructions and agree—even if language is a crucial component in well-functioning human cooperation (Smith 2010). The central point is instead what the replicator does *for itself*.

Even when we fight we cooperate in the use of language. 'English'—the communication system itself—is favored in all interactions using English, even if our genes are not. Language can thus be an important explanation of the universal cooperation between humans—the cooperation is not so much about us or our genes anymore, it is about language and other cultural traits cooperating with themselves.

As we saw above, language has both a biological and a cultural component. We have, for example, a number of purely biological adaptations that enable us to talk and understand, ranging from fine motor control of breathing to vocal cords and parts of the brain that are specialized for language production and understanding.

Once these biological abilities are in place, the cultural components of language are free to evolve separately from the genes. For example, you can learn Mandarin or Quechua even if you are from Iran. There is no gene determining what languages you can learn.

This means that the replicator 'meme' has to be understood separately from the replicator 'gene'. Biology and culture are potentially completely separate and free to evolve according to their own rules, although, as we shall see, this is definitely not true for all cultural traits.

Perhaps, then, it is not our own or our genes' cooperation that is the key to an understanding of the exaggerated cooperation between humans. Could our exaggerated cooperation be a product of traits cooperating with each other 'over our heads'?

This does not imply consciously chosen cooperation between 'memes', but cooperation as in 'collective action' in the same manner as the purely mechanistic 'collective action' of genes. If you remember, *Cooperation is the collective functioning of some kind of units for the benefit of themselves and/or their component parts.*

Now, the collectively functioning units are 'memes' and the benefit may be only for their own benefit. As we shall see, such cooperation can sometimes even be to the detriment of our genes. Has language transformed us from active agents to something more akin to a well-fertilized soil bed for a completely new set of replicators?

Because surely you didn't decide freely what language to speak this morning, or what culture to grow up in? When you got dressed, did you choose between a suit, a Baroque dress, some hides, two loincloths, a toga and some penis sheaths? These are all clothes that have been used by humans, yet are not 'available' as options. Could you go outside naked if you wanted? Could you just as well speak Hindi?

We do not have much say in the design of our nature—our bodies and their abilities are 'given' to us. The same is true for the design of our culture. How much of what you think have you actually decided to think about yourself?

It's time to look more closely at human culture.

References

Arensburg, B., Schepartz, L. A., Tillier, A. M., Vandermeersch, B., & Rak, Y. (1990). A reappraisal of the anatomical basis for speech in middle Paleolithic hominids. *American Journal of Physiological Anthropology, 83*, 137–146.

Berwick, R. C., & Chomsky, N. (2016). *Why only us: Language and evolution*. Cambridge: MIT Press.

Blackmore, S. (1999). *The meme machine*. Oxford: Oxford University Press.

Bower, B. (1990). Getting out from number one: Selfishness may not dominate human behavior. *Science News, 137*, 266–267.

Cheney, D. L., & Seyfarth, R. M. (1990). *How monkeys see the world*. Chicago: University of Chicago Press.

Chomsky, N. (1957). *Syntactic structures*. The Hague/Paris: Mouton.

Dawes, R. M., van der Kragt, A., & Orbell, J. M. (1990). Cooperation for the benefit of us—not me, or my conscience. In J. Mansbridge (Ed.), *Beyond Self-Interest* (pp. 97–110). Chicago: Chicago University Press.

Dawkins, R. (1976). *The selfish gene*. Oxford: Oxford University Press.

Dawkins, R., & Krebs, J. R. (1978). Animal signals: Information or manipulation? In J. R. Krebs & N. B. Davies (Eds.), *Behavioural ecology* (pp. 2823–09). Oxford: Blackwell Scientific Publications.

Deacon, T. W. (1997). *The symbolic species: The co-evolution of language and the human brain*. Allen Lane: The Penguin Press.

Dornhaus, A., & Chittka, L. (2014). Why do honey bees dance? *Behavioral Ecology and Sociobiology, 55*, 395–401.

Ettlinger, M. (2014). *In what sense is the word recursion' used in linguistics*? https://www.quora.com/In-what-sense-is-the-term-recursion-used-in-linguistics

Fitch, W. T. (2010). *The evolution of language*. Cambridge: Cambridge University Press.

Gardner, S., & West, S. A. (2010). Greenbeards. *Evolution, 64*, 25–38.

Ghirlanda, S., Lind, J., & Enquist, M. (2017). Memory for stimulus sequences: A divide between humans and other animals? Royal Society Open Science, *in press*.

Grüter, C., Balbuena, M., & Farina, W. (2008). Informational conflicts created by the waggle dance. *Proceedings of the Royal Society, Series B: Biological Sciences, 275*, 1321–1327.

Gärdenfors, P. (2000). *När Homo blev sapiens*. Nora: Nya Doxa.

Henrich, J. (2016). *The secret of our success: How culture is driving human evolution, domesticating our species, and making us smarter*. Princeton: Princeton University Press.

Herrmann, E., Call, J., Lloreda, M., Hare, B., & Tomasello, M. (2007). Humans have evolved specialized skills of social cognition: The cultural intelligence hypothesis. *Science, 317*, 1360–1366.

Hofstadter, D. R. (1979). *Gödel, escher, bach: An eternal golden braid*. London: Penguin.

Kay, R. F., Cartmill, M., & Balow, M. (1998). The hypoglossal canal and the origin of human vocal behavior. *Proceedings of the National Academy of Sciences, 95*, 5417–5419.

Krause, J., Lalueza-Fox, C., Orlando, L., Enard, W., Green, R. E., Burbano, H. A., et al. (2007). The derived FOXP2 variant of modern humans was shared with Neandertals. *Current Biology, 17*, 1908–191.

Lindenfors, P. (2013). The green beards of language. *Ecology and Evolution, 3*, 1104–1112.

Ostrom, E., Walker, J., & Garnder, R. (1992). Covenants with and without a sword: Self-governance is possible. *American Political Science Review, 86*, 404–417.

Parkvall, M. (2009). *Lagom finns bara i Sverige och andra myter om språk*. Stockholm: Telegram Bokförlag AB.

Pilley, J. W., & Reid, A. K. (2011). Border collie comprehends object names as verbal referents. *Behavioural Processes, 86*, 184–195.

Ridley, M. (2009). Matt Ridley talk at Thinking Digital 2009. https://vimeo.com/7536594

Rainey, H. J., Zuberbühler, K., & Slater, P. J. B. (2003). Hornbills can distinguish between primate alarm calls. *Proceedings of the Royal Society, Series B: Biological Sciences, 271*, 755–759.
Számadó, S., & Szathmáry, E. (2006). Selective scenarios for the emergence of natural language. *Trends in Ecology & Evolution, 21*, 555–561.
Smith, E. A. (2010). Communication and collective action: Language and the evolution of human cooperation. *Evolution and Human Behavior*, 31, 231–245.
von Frisch, K. (1967). *The dance language and orientation of bees*. Cambridge: The Belknap Press of Harvard University Press.
Wilde O 1891 *The Critic As Artist*.
Wu, J., Balliet, D., & Van Lange, P. A. M. (2016). Gossip versus punishment: The efficiency of reputation to promote and maintain cooperation. *Scientific Reports, 6*, 23919.
Wynne, C. (2007). Aping language: a skeptical analysis of the evidence for nonhuman primate language. *eSkeptic*, October 31.

The Last Piece of the Puzzle?—Cooperation Over Our Heads

Abstract Human evolution and the evolution of cultural capacity are discussed. The process of cultural evolution is described. Cultural evolutionary explanations of the "problematic" parts of human cooperation are outlined. Current (and somewhat disputed) knowledge around how humans can be seen as nodes of a network and how traits spread in such networks are presented. Memetic selection is discussed.

> Every boat is copied from another boat… Let's reason as follows in the manner of Darwin. It is clear that a very badly made boat will end up at the bottom after one or two voyages, and thus never be copied… One could then say, with complete rigor, that it is the sea herself who fashions the boats, choosing those which function and destroying the others.
>
> Émile Chartier: Propos d'un Normand (1908; as quoted in Rogers and Ehrlich 2008)

The human capacity for culture demarcates humans from other animals. Animals can learn things too, but not to the degree that we can—and must; humans are addicted to culture.

However, even the simplest of animals can learn things—learning doesn't even required a brain, not even multicellularity. The one-celled organism *Paramecium caudatum* can be taught to prefer either light or darkness through the use of electric shocks. Oddly, *Paramecium* cells prefer shocks and move towards the condition—light or darkness—they associate with these (Armus et al. 2006). This preference is then, in some yet unknown way, biochemically encoded into the cell itself—chemically learnt.

(This could perhaps come as some consolation to tired teachers; even organisms without brains can learn—all that is needed are some electric shocks.)

However, most organisms where learning has been studied are animals with brains. Brains govern body functions and react to external stimuli transmitted via sensory organs. Animals can be taught to find food using similar associations as the *Paramecium* cells (minus electric shocks—the important thing is to use something the animal regards as 'pleasant').

Animals easily learn to associate a food outlet with some sort of signal, for example light or sound. If the animal is then consistently attracted towards that light

P. Lindenfors, *For Whose Benefit?*, DOI 10.1007/978-3-319-50874-0_9

or sound, learning has occurred. This form of learning, to associate a signal (light, sound, smell) with a phenomenon (food, danger, partner) is termed *conditioning* and is extremely common and something all animals can do.

However, some animals can also learn in other ways than through their own direct experience: they learn from other individuals. An experiment on birds clearly shows this. Two caged birds are arranged so that the birds can see each other, but not what is above the other's cage.

If you let one bird see the silhouette of a raptor it gives a warning call, which understandably worries the other bird, even though it can't see the silhouette. If you show that bird another silhouette, for example of a bottle, the bird will learn to associate bottles with danger. Not because of any bad experience of bottles, but from learning from the other bird's reaction.

This is a common form of social learning (Bonner 1980). What the bird has learnt isn't a new behavior. It has learnt to associate an already existing (genetically inherited) behavioral complex (what to do in danger) with an object (bottle silhouettes). In the wild it would have learnt danger in the same way, but made the correct association with raptors.

There are whole systems of such knowledge—learnt associations modifying genetically influenced behaviors—transmitted from generation to generation, not via genes but through social learning. For example, some birds and whales transmit song in this way. The same applies to food habits and paths of migratory birds, associations often learnt from other individuals.

In biology, the transmission of knowledge between individuals is termed *cultural transmission*. This is also a form of inheritance; inheritance transmitted from individual to individual, from generation to generation, but not via DNA. All learning isn't cultural transmission, however, only social learning—learning from others.

But are bird song and migratory routes *culture*? If you define 'culture' as any trait that can be inherited through learning from other individuals, then yes. By this definition, traits such as bird song, whale song and migratory routes are the simplest forms of culture.

Using this definition it is also possible to analyze cultural differences between animals. Songbirds that learn from each other, for example, don't all end up with the same version of song. In northern Sweden, blackbirds sing somewhat differently from blackbirds in southern Sweden. They learn dialects. If you are obsessive enough as an ornithologist you can learn to hear the blackbird's origin.

Another example: on Zanzibar there are a group of Colobus monkeys that live on leaves from Indian almond and mango trees. These leaves have a high concentration of fenol, a substance that is poisonous and can disturb congestion, presumably to deter leaf eaters. However, the Colobus monkeys have discovered that they can eat the leaves—that are very nutritious—if they also ingest charcoal from old fires. Through this learnt trait—transmitted from mother to offspring—the group has achieved a much higher population density than their conspecifics that haven't learnt this behavior (Struhsaker et al. 1997).

Note that this has very little to do with the biological evolution of new traits since there is no genetic coding for self-medication through charcoal eating. The two species of tree are too new to Zanzibar for this to have happened. Instead this is a form of proto-culture (Dennett 2003).[1]

The most well documented form of culture in animals is tool use among chimpanzees. In some parts of Africa Chimpanzees can, for example, put a nut on a large rock and then crush its shell using a smaller stone. However, only certain chimpanzees have learnt this trick.

In some populations the behavior is missing because there are no nuts in their area. In other populations there might be nuts, and the necessary tools lie strewn across the landscape, but no chimp has figured out what to do. In particular populations, chimps have learnt to expertly fish for termites by sticking sticks into termites' nests; in other populations no one has yet had the brainwave.

Researchers have identified 38 behaviors that exist in some populations of chimpanzees but not in others and that cannot be explained by differences in environment (Whiten et al. 1999). Together with a colleague—Johan Lind—I have published results indicating that most likely, the females are the carriers of this cultural information as they are the ones that move between populations and the ones from whom infants learn (Lind and Lindenfors 2010).

The 38 socially transmitted behaviors in chimpanzees are the pinnacle of animal culture. There is no comparable cultural diversity in other animals. This is *astonishingly* less complex than cultural diversity in humans. Computer models indicate that it in some situations is much more beneficial for survival and reproduction to learn from others than to learn by yourself (Rendell et al. 2010). So why don't other animals than us have cultural evolution?

We humans build incredible buildings, use fantastically complex machinery, live in gigantic societies with intricate formalized as well as unspoken rules for how to behave and live. Humans have invented things such as atonal music and abstract art. With the aid of cultural knowledge, humans have colonized the entire earth. No other animal has dominated the world in the way humans do presently.

None of this is due to some marvelous human biological adaptation, but it is all due to our cultural abilities. For example, cultural adaptations make it possible for Inuit to live in the extreme polar landscape. The Inuit haven't evolved fur as a biological adaptation, but have instead learnt how to use other animals' furs to keep warm. Our collective human success is *completely* due to culture.

[1]It is theoretically possible that the behavior will get coded into the genome eventually, given enough time, in the same manner as some believe that language-learning ability has been encoded in human genes over a long time period. For example, a mutation that gave the Colobus monkeys a mutation which meant that they the experienced the charcoal as tasty would give a quicker access to the poisonous leaves for young monkeys because they wouldn't have to go through a learning process. This possibility of learnt behaviors being written into a genome is termed the Baldwin effect.

What seems to be missing in animals, what we can do but they can't, is to steadily accumulate novel knowledge, to add culturally learnt traits together to create new culture.

Isn't this ability a biological ability? There are biologists who argue that culture shouldn't be viewed separately from biology, but be understood as a novel kind of biological trait. American anthropologist John Tooby and his wife, evolutionary psychologist Leda Cosmides, have for example written that 'Human brains, human behaviors, human artifacts and human culture are all biological phenomena' (Tooby and Cosmides 1992). As we shall see, this is a very limited view of culture.

That the *ability* to accumulate cultural knowledge has a biological foundation, as we saw in the last chapter, is not the point. Humans have an unprecedented ability to learn from others through language, better learning mechanisms, better memory, sequence processing, thought, evaluation, planning and inventiveness than any other animal. But the really exceptional thing with humans is not this collection of traits, but what they enable: cultural evolution.

Two geneticists, the Australian-American Marcus Feldman and the Italian Luigi Luca Cavalli-Sforza, published in 1976 the first dynamic model of how genes and culture can coevolve. The American biologists Charles Lumsden and E.O. Wilson attempted to build on these insights 1981 (Wilson and Lumsden 1981), but were accused of exaggerating the importance of genes in cultural evolution.

Instead, Cavalli-Sforza and Feldman themselves followed up their initial model in 1981 (Cavalli-Sforza and Feldman 1981), but it wasn't until 1985 when the anthropologist Robert Boyd and the biologist Peter Richerson, both Americans, published their book *Culture and the Evolutionary Process* that modern research on cultural evolution really started (Boyd and Richerson 1985). The insight all these authors argued for, but that was made exceptionally clear by Boyd and Richerson, was that we have two forms of inheritance that both evolve—*dual inheritance theory*—something we shall investigate more closely, after a brief review of the archaeological record to see what it tells us about how and when cultural capacities evolved.

A Slow History

Human adaptations for cultural evolution are young in a biological perspective—but all humans have them. Put another way, you can take a human from any place in the world and put them in an American school and the children will become fully American in their way to be and will learn all that American children normally learn. Concerning this ability for culture, humans are astonishingly alike.

As I and several colleagues have pointed out, this observation alone means that the ability for cultural evolution must have been with us at least since our common origin in Africa, at least 170,000 years ago when the human family tree branched off for the first time (Lind et al. 2013). Had this not been the case, the *same* mutations for cultural ability would have to have occurred several times, a highly unlikely form of parallel evolution. The consequence of this common origin of cultural capacities is that we also should be able to take a child from the dawn of *Homo sapiens* and put her in a contemporary American school without noticing any major difference in abilities between her and other children.

But how do we explain this: If humans have possessed the same abilities for at least 170,000 years, why did such things as farming, writing, and technology not all just pop into existence immediately? Why has all this development occurred the last 10,000 years? And don't we find archaeological evidence for culture that is much older than this?

Many animals use tools, but the oldest examples of *manufactured* tools that we know of are the so-called Oldowan tool, simply fashioned stone tools from about 2.6 million years ago. [Though it has recently been reported that capuchin monkeys problematically enough smash stones and thus produce similar artefacts, probably to get at powdered quartz or lichens (Proffitt et al. 2016).]

Non-manufactured stone tools are used by for example chimpanzees and sea otters, so they are not useful as a sign of the presence of human culture. 2.6 million years is so long ago that it isn't even modern humans we are talking about, but *Homo habilis* and *Homo ergaster* (and possibly late *Australopithecus garhi*, but this is contested). The Oldowan method of making stone tools lasted for almost a million years: from 2.6 to 1.7 million years ago.

A million years scarce improvement or substantial change!

After this period the Oldowan tools were replaced by Achulean tools. The difference between the two types is that the earlier tools were worked only once to produce a desired shape, while the Achulean tools were worked a second time, this time with tools that were not only other stones but made of bone, horns, or wood. This time it was *Homo erectus* who made the tools, a species of human that spread from Africa to Europe and Asia, so this type of tools are found everywhere where they lived. This period of stone tool manufacture lasted for more than a million years: from 1.8 million to 100,000 years ago.

More than a million years again—on three continents—again with scarce improvements! Why didn't techniques improve?

During this period we see, however, shifting complexities in tool collections, some indicating more advanced traditions, some indicating less advanced traditions—as if there are fits and starts of tool-making competence. These traces are the first indications of tool making improvements, indicating the rudiments of cumulative cultural evolution. As tools become more and more important, natural selection would have favored individuals who are better cultural learners—through the coevolution of genes and culture described by dual inheritance theory—but as we shall see, cultural progress also depends on population size and density (Henrich 2016).

The next type of stone tools, the Mousterian, was different. Now, the stone tools were fashioned from a core—a larger stone—from which smaller sharp stones were chipped away. This type is associated with the Neanderthals (*Homo neanderthalensis*) and was produced for 270,000 years: from about 300,000 years ago until 30,000 years ago.

Even if the exact timings of these events can change as new finds are made, the general pattern is crystal clear: technological development was painfully slow—practically nonexistent—in early human history (Boyd and Silk 2009).

But why? Once you have thought of how to make tools by chipping pieces from larger stones, why doesn't somebody refine the method at all—for a million years? And why does every major step in stone tool evolution seem to be associated with a new species of human?

What this suggests is that these human-like creatures just couldn't do any better. A chimpanzee seems unable to learn language. Early humans seems to have had a similar kind of constraint concerning better stone tools.

The size of the human brain increased markedly during this time, from late *Australopithecus* 2.5 million years ago to anatomically moderns *H. sapiens sapiens*, especially during the time of *H. erectus*. Our current brain size was arrived at just 200,000 years ago (to be picky, the brain was then even a bit bigger then than it is now). This indicates that the non-existent stone tool evolution can at least partly have been due to limitations in cognitive capacity.

Our brains are so ridiculously large in relation to our bodies that before the advent of modern medicine it was a major health hazard for women to give birth—this despite the fact that our brains have higher cortical folding and is more densely packed than other primates' brains. Even though we are born with such a large brain, there is a lot of further growth after birth—eventually the brain triples in size from birth size while other apes' brains only double.

Moreover, the myelination of the brain—a process that makes the brain more efficient but less formable, less able to learn—differs from that of apes. Chimps have 15% of their cortex already myelinized at birth, humans only 1.6%. While this process finishes at reaching adulthood in chimps, humans are only 65% complete by then—we continue brain development until our late twenties (Henrich 2016).

Perhaps there was some form of co-evolution between tool use and the evolution of the brain during this time (Myrdal 2004–2005), or between social competence and the evolution of the brain (Wills 1993). Or perhaps between language and the brain? (Deacon 1997) The fossils provide only few clues.

It is somewhat clear that specific methods of tool-making are associated with specific species of (pre-) humans in the archaeological material. Early stone tools should therefore perhaps better be regarded as expressions of genes—phenotypes occurring outside the body. Richard Dawkins has named this type of external genetic traits *extended phenotypes* (your *phenotype* is 'what you are like' and is determined by your *genotype,* which is your personal collection of genes) (Dawkins 1989).

There are many examples of extended phenotypes—phenotypes outside physical bodies—in nature. Beaver dams, birds' nests, termite mounds and anthills are all products of genetically programmed behaviors and have their expression outside the bodies of the animals. Wagtails build their nests out of grass and other plants, magpies build nests out of twigs, mud and feathers, while weaverbirds weave intricate basket-like nests that are specific to each species of weaver bird. These are all complex constructions with many separate subsections, but they are not cultural traits; these building processes are nothing the birds learn, but something they already 'know'.

The slow cultural evolution in human pre-history is therefore not really strange. From a biological perspective, slow gradual evolution of extended phenotypes is the normal, what is to be expected.

What about the Neanderthals? Neanderthals made advanced stone tools, used fire in a controlled manner and even seem to have buried their dead, but something seems to be missing: they never progressed from the most primitive stone-age living conditions. It is not known if this was because they couldn't, or because they died out before proving themselves, or because living conditions kept the population density down. Maybe we will never know.

Fossil finds indicate that primitive cultural competencies hay have evolved in *H. erectus*, and actually to a large degree were present in Neanderthals. In *H. erectus*, stone tools seem to have been refined, then these developments were forgotten, then invented again, and so on. Just as would be expected in a species that is developing true cultural competencies but is not quite there yet, or alternatively, where the population is too small or keeps crashing (Henrich 2016).

But approximately 200,000 years ago a new species of human entered the scene: *H. sapiens*, the species to which we all belong. *Now everything changed*. With the entrance of modern humans on the evolutionary stage, technology for the first time started to evolve more quickly than biology.

In South Africa have been found 160,000-year-old traces of a more advanced stone-age culture: small sharpened stones that were probably used as arrowheads. The same people also used red ochre, maybe for decoration purposes. In caves in Morocco researchers have found Nassarius shells painted with red ochre that are between 120,000 and 80,000 years old. The caves lie approximately 40 km from the coastline, so the shells had been carried all that distance. Other similar, pierced shells (ornaments?) have been found in Algeria, two hundred kilometers from the coast. Together with finds of obsidian in Ethiopia and other parts of East Africa—far from their origin—these shells are the first signs of trade.

The British science journalist Matt Ridley has hyped up these early signs of trade to suggest that trade is what really got human cultural evolution started by enabling specialization (Ridley 2010a, b). As we saw in the first chapters, specialization is a central part of our understanding of when cooperation leads to some unit which is larger than just the sum of its parts. Certainly, trade and specialization of tasks are uniquely human capabilities, but it is hard to imagine trade without language and trinkets to trade with each other, so trade is more likely a consequence than a cause of human culture.

Whatever this causal association turns out to be, trade is one of many signs of a developed capability for culture that is unique in the animal kingdom. Cultural evolution in *H. sapiens* occurred faster than in any previous human species, Neanderthals included. 45,000 years ago, people in the western part of the Eurasian landmass could fashion 'knife blades' by chipping them from stone cores. A 35,000 year old bone flute is the oldest musical instrument ever found. In layers that are 24,000 years old, archaeologists have found bone points for spears, and

26,000 years ago humans learnt to make needles. Spear-throwers out of bone—so called atl-atl—are found in layers from 18,000 years ago. Not much later humans invented the bow and arrow. Just as old are finds in Ukraine of ornaments made of shells from the Black Sea as well as amber from the Baltic Sea.

Methods for making tools were still slow in developing throughout this time, the big brain notwithstanding. Stone tools are, however, just the tip of the iceberg since it is much easier to make tools from other materials such as wood and plant parts, as well as leather, bones and other animal parts. Because these decompose, however, only stone tools remain.

From Ötzi, the iceman that was found 1991 frozen in a glacier in the Italian Alps, we have gained a more complete picture of stone-age technology. Ötzi lived 5300 years ago, at the turn of the stone-age and the bronze-age.

He had a fur coat, sewn by at least two persons, where the original stiches were nicely sewn with sinew but where all repairs were hastily done with grass thread. On his head he wore a fur hat and closest to the body a loincloth that functioned as underwear. Ötzi's legs were warmed by leggings of animal hide and his shoes were lined with grass and so nicely done that some researchers have come to the conclusion that specialized shoemakers must have existed at the time.

Ötzi also carried a case containing flint, a flint-bladed dagger fitted to an ash handle, and a quiver with 14 viburnum and dogwood-shafted arrows. Only two of the arrows were ready to be used, with flint tips and fletching, while the others were in various states of repair or manufacture. What must have been his prize possession was a nicely cast and forged copper axe perfectly fitted to a handle made of yew.

Ötzi had two baskets made of birch bark, one of which contained a glowing ember in a package of fresh maple leaves, but he also carried flint and pyrite for creating sparks as well as a type of tinder fungus. Interestingly enough, he also carried birch fungus on a string—birch fungus is known to have anti-inflammatory and anti-bacterial properties. An X-ray of his skeleton indicates that he must have suffered from age-conditioned or strain-induced degeneration in the knees, ankles and lumbar spine. On some of these places he had tattoos, indicating attempted treatment. Both Ötzi's clothes and his baggage indicate the existence of specialized artisans and trade (Ridley 1996; South Tyrol Museum of Archaeology web-site).

But even if technological progress was quicker for *Homos sapiens* than ever before, there are still thousands of years between the technological advances. Despite the fact that *H. sapiens* early (from the beginning?) probably had the mental capacity for magnificent cultural progress, cultural evolution progressed only slowly. Why?

Think about it this way. Contemplate, for a moment, what in your own arsenal of cultural knowledge that you have learnt by yourself—*all* by yourself.

Language? Not at all, that is something you have picked up from others. What can be eaten and drunk? Possibly some exotic forest species, but otherwise our diets are pretty uniformly culturally determined. How to make stone axes? That the sun is a star and the earth round? That Napoleon was a French emperor? The law of gravity, theory of relativity or the speed of light in a vacuum? That Muhammed is

God's last prophet, Jesus the savior of the world, or that Krishna was a god that came to earth to enlighten us in questions of right and wrong?

If you think about it, is there *anything* that you have leant *all by yourself* through a personal interaction with the environment? I am referring to *knowledge*, not a nature given proficiency like being able to breathe, walk, throw rocks or interpret shifting neuronal impulses across your retina as an approaching predator. Most probably, you won't be able to think of anything (or you misunderstood the question). Pretty much everything we know is knowledge we have learnt from someone else.

The American historian George Basalla draws the logical conclusion from this and writes in his book about technological evolution that: 'Any new thing that appears in the made world is based on some object already in existence. (...) each new technological system emerges from an antecedent system, just as each new discrete artifact emerges from antecedent artifacts' (Basalla 1988). But surely something must have been the first object? In these cases, Basalla shows, these objects are natural.

This puts a natural speed limit on how fast culture can emerge and evolve. It takes time to build common knowledge when all new knowledge has to build on what already exists. Culture is a collective endeavor.

Note the similarity between language and culture in this view. What separates human language from the symbolic communication that animals can learn is that we can construct language from other language—this is why we can have lexicons. We have passed the symbolic threshold and don't need to anchor language in external objects (Deacon 1997).

Culture works similarly. We can construct culture from other culture without having to involve reality, yet another reason to believe that language was a key condition for human culture to be possible.

The Austrian-British philosopher Ludwig Wittgenstein expressed it clearly 'We may say that thinking is essentially the activity of operating with signs' (Wittgenstein 1958). Not that animals can't think, or that *all* thinking is symbol manipulation, but the specific thinking that makes culture possible is based on the ability to express new ideas out of old; new symbols out of combinations of other symbols.

We can all contribute something to the accumulated set of human knowledge, but in a larger perspective, human knowledge—culture—is something that *societies* have and transmit across generations. Our success as a species is not due to individual capacities, but to our collective capacities—cultural knowledge is communal.

500 years ago there were incredible cultures in Europe and Asia. But in Australia, in isolated Native American tribes in the Americas, and in certain African ethnic groups, humans were still using stone-age technology, mainly because of geographic circumstances that resulted in smaller populations and a lower population density (Diamond 1997).

Fewer people means fewer opportunities to invent, transmit and preserve culture. In studies of islands in the Pacific Ocean this has become clear. Islands with large populations have more tools and greater cultural complexity than islands with small populations (Kline and Boyd 2010).

The most dramatic illustration of the effects of low population size is from Tasmania, just before the English killed off the native population. The Tasmanian aborigines became separated from the mainland aborigines when the ocean level rose approximately 8000 years ago. Archaeological finds have shown that the Tasmanians also stopped trading with the aborigines on the Australian mainland after the separation, so the population preserving the cultural knowledge of the Tasmanians was very small for a long time.

When the British came to Tasmania in 1803 they found between two- and eight thousand people who had forgotten how to make fire. Appointed men carried glowing embers from camp to camp. The inhabitants couldn't make fishing tools anymore and had stopped eating fish completely—somewhat odd for an island population. They couldn't manufacture tools out of bone or clothes to protect them from the cold. They had no shafted tools and no barbs on their spears. Archaeological finds show that they previously knew all these things.

This loss of knowledge has nothing to do with genetics; the Tasmanians were simply too few to maintain their collective knowledge. Of course there are differences between individuals in what talents and capacities we are born with, but no apparent systematic differences in mental abilities seem to exist between different human groups to explain why we in the West were the first to master modern technology (Diamond 1997). The cultural differences are not explicable by differences in mental abilities.

Think of humanity's technological journey over the past century. No-one believes that our genes have changed substantially during this time. There has simply not been a large enough number of generations for 'genes for' the use of computers or the driving of cars to emerge and spread over humanity. On the contrary, (almost) everyone seems able to learn to use a computer or drive a car, if you just start early enough.

The explanations of our progress *must* therefore lie in culture itself, how it has evolved over time.

Cultural Evolution

Culture—in the context used here—means any trait that we can learn from others. This learning gives rise to a type of *inheritance* distinct from genetic inheritance. In contrast to genes, cultural inheritance can run between any two individuals. We don't learn only from our parents but also from our children, or from any individual at all. We learn from watching animals, or through reading books.

There is also *variation* between different cultural traits. Some people speak French while others speak Swedish, some plow with a plow while others use pick and shovel, some like Chinese opera while others prefer rock. There are many cultural traits that we can copy from others.

But not all of these traits have an equally large chance of being copied by others. For most people, it is much more attractive to copy a behavior that is about satisfying a craving through consumption and gratification than one that is about celibacy and ascetics, simply because the former is more enjoyable than the latter. It is also *easier* to learn certain explanations than others. For example, it is easier to learn that 'Brahma created the world with all organisms as we find them today' than to spend years and years in school to understand the intricate details of the theory of evolution.

You may also more easily copy cultural traits from people of high status, or from those who you deem to have succeeded in life, or from the majority. All cultural traits simply don't get copied equally—there is an imbalance in the transmission of cultural traits (Richerson et al. 2003).

So there is *inheritance*, *variation* and *an unequal transmission of traits*—the same three factors that are necessary for biological evolution. Is cultural change therefore an evolutionary process?

This is not a new thought. In fact, the word 'evolution' was used to describe cultural change even before Darwin borrowed the term to biology. The problem is that 'cultural evolution' was coined by Herbert Spencer.

British philosopher Herbert Spencer, one of the founders of sociology, was overjoyed when Darwin published *On the Origin of Species*—a scientific description of the mechanisms behind change in living organisms. At the time, Spencer saw evolution in *all* processes of change, meaning a slow change/improvement from one condition to another. When Spencer read Darwin he realized that Darwin's more formalized theory should also be applicable to culture. Spencer saw that one of the creative forces in the evolutionary process was that less adapted versions got weeded out. He took all of this thinking and applied it to humans.

With this, Social Darwinism was started—the thought that less adapted humans and cultures have to disappear if progress is to be possible—and that this disappearance is something good. Some Social Darwinists even believed that we can speed up progress by eliminating 'less capable' humans and cultures; a kind of breeding-project, but on humans and cultures (Lewens 2013).

This is not a particularly pleasant theory and also a kind of misunderstanding of evolutionary biology. It is rare that 'kill them all' is the best evolutionary solution in biological evolution. It doesn't have to be the best killer that is the best adapted, it can equally well be that the individual who is best at nurturing offspring, or invests the least in muscle mass, or—of course—the one cooperating the best, that is most adapted.

Because of Social Darwinism, 'cultural evolution' long had bad publicity. This doesn't mean that there was no evolutionary thinking in studies of cultural change. On the contrary, the interest in the topic has waxed and waned in periods since Darwin and Spencer.

For a long time, the different 'stages' of societies was an important topic; the stages that societies went through from group to tribe to chiefdom to state. There were many discussions of exactly what stages that had to be included in the analyses and if each population really had to pass through all. There were also discussions on whether cultural evolution means that all societies pass through the same stages in their evolution or to what degree societies can evolve along their own unique trajectories.

If societies change along the same trajectories one immediately makes the association with notions of primitive and advanced. Are non-Western societies barbaric and just haven't evolved as far as we have? Or are all societies unique travelers in the space of different possible cultural systems each of which has found its own optimal solution of how to organize? (Carneiro 2003).

Modern research on cultural evolution has mostly left stage thinking behind. It is not considered as important anymore, whether societies evolve on their own or along similar evolutionary trajectories. Instead, the focus has shifted to the cultural traits themselves. Questions studied include how cultural traits spread, what types of cultural traits that become more common and why, if traits within different cultural domains (for example, music, farming, religion) change in similar or different ways, and if certain cultural traits evolve together and—if yes—why.

If culture works approximately according to the same principles as genetic inheritance, could methods be transferrable from biological to cultural evolution? Two inheritances—one theory, one set of methods! This is what some researchers working on cultural evolutionary theory have attempted to do under the heading *Dual inheritance Theory*.

Approaches include modeling trait transmission and constructing trees describing relatedness between languages and cultural traits in the same way as one builds relatedness trees (phylogenies) over a group of organisms. But is it really possible to build a phylogeny over Inuit methods of canoe-building? Or of sword evolution across the ages? Yes and no. The project to move biological theories to analyses of cultural change has met with mixed success because of a number of fundamental differences between biology and culture. These differences all stem from one crucial difference: that between genes and 'memes'.

Biological traits are inherited through particles—genes—that are materially transmitted from one individual (parent cell/individual) to another (offspring cell/individual). These transmitted particles contain instructions for biological traits. Contrast this to how learning takes place. In learning, nothing material is transmitted. To circumvent this problem it has been suggested that cultural transmission be viewed *as if* something material is being transmitted. Richard Dawkins has even suggested a name for these thought 'particles': *memes* (Dawkins 1976).

Once a name has been established for a hypothetical particle it suddenly feels as if you know what you are talking about. But there are obvious problems. Foremost, there exist no cultural 'particles' as there does for genes. But also, when you learn that hovercraft in Swedish is called 'svävare' you may have picked up a meme. But when you learn that the god Shiva maintains the world through meditation, how many cultural particles have you picked up? Or when you learn agriculture? What is the smallest indivisible unit of culture?

If you examine the problem more closely it becomes apparent that the concept of 'meme' is indefinable. There are researchers that claim that the same difficulties apply to genes, but this is not really true, you can very well imagine genes as particles of inheritance and use this notion to calculate spread (Laland and Brown 2002). It is even possible to use biochemical methods to measure the prevalence of different genes in populations of organisms.

Another problem with 'memes' is that the distinction between an instruction for a trait and the trait itself isn't crystal clear. In biology, this distinction is fundamental; it is why acquired traits cannot be inherited, which is not strictly true for cultural traits. But that really depends on the level of analysis. If you take the view—which some researchers do—that cultural transmission isn't so much about individual inheritance, but more a form of contagious 'parasite' where ideas can 'infect' one person from another, then you can no more inherit acquired traits than you can parasites. Or can you? Someone can give you a stone axe, after all.

Richard Dawkins has formulated the problem with our current understanding of what a 'meme' is: 'Memes have not yet found their Watson and Crick; they lack even their Mendel' (Gleick 2011).

Because of the more 'fluid' inheritance of memes in that they are not transmitted materially, cultural information can be transferred between any two individuals. Not only can a child learn from her parents, but parents can learn from their children. If you like, you can travel to South America to learn to manufacture blowpipes from Native Americans.

This possibility of free movement of ideas makes certain units that are taken for granted in biology dissolve into thin air. In biology, an *individual* is (almost always) someone who has received a certain set of genes that are then carried through life. A *cultural individual*, on the other hand, can change the traits received—you can learn throughout life, just as you can get new parasites through life. Even if you have had the flu it isn't part of your person.

The same is true for a *species* which in biology is a group of individuals who can have fertile offspring with each other. This is non-existent in cultural evolutionary research. For example, you cannot count ethnicities or countries as cultural units since there is a lot of cultural information that seamlessly crosses borders. And it doesn't work to define 'cultural species' from language since you can learn things also from people not speaking your language—and language elements are borrowed between languages all the time.

Another consequence of the non-material, non-particulate transfer is that cultural information is processed when transmitted. If you have an idea you want to get across you have to 'translate' it from impulses in your own brain into for example spoken or written language. In comparison, genes are transmitted directly as material strings of information. The recipient of cultural information must in turn 'translate' the language to impulses in his own brain. Things you learn then rebuild the brain—this is how learning happens. Connections between different nerve cells become strengthened of weakened. Genes in the nerve cells are turned on or off. Knowledge becomes a part of your physical self despite the fact that no material inheritance has been transmitted.

Inside its new home, the new brain, the learned trait meets with a novel environment of other traits and proficiencies not present in the source brain. Brains are already inhabited by ideas, but there are different ideas in different brains. There are thus no guarantees whatsoever that the same information is interpreted similarly by the two brains. On the contrary, it is almost certainly interpreted differently, as the encoding of knowledge in a brain during learning will occur differently in the two brains.

It can be difficult for two conflicting cultural traits to co-inhabit the same brain. The conflict that ensues—when, for example, circumstances force you to act in conflict with some deeply felt moral conviction, or when a prophesied end of the world doesn't happen—in psychology goes by the term 'cognitive dissonance'. The brain (or rather, your 'self') in some manner tries to evade these kinds of contradictions. Thoughts have to cooperate (function together) too.

No cultural information is 'given' so that it exists in some permanent, unified form that can be adopted. For biological traits, all genetic material that an individual needs is transmitted at conception materially—encoded in actual physical material. The individual is at conception complete concerning biological inheritance—but totally empty concerning the cultural inheritance.

All cultural information has to be adopted subsequently and be interpreted by and stored in the brain. This means that culture really isn't something you adopt, but something that has to be created in each of us—cultural reality is cooperatively created in all our brains.

This is how the talk about reality being a cultural construct should be understood. Not that no reality exists outside ourselves, but that *cultural reality*—our common knowledge and understanding of the world—exists only as common 'agreements' (that can fit actual reality more or less well). Culture is a group phenomenon—a cooperative phenomenon.

Think of it this way: there is no guarantee that information transmitted between brains is honest or true. 'A spirit inhabits the tree over there' is the exact same type of information transfer as 'There is a lion besides the large collection of boulders'. As long as the opposite hasn't been confirmed you may be wise to avoid both tree and boulder collection. The existence of a spirit in a tree can in this way become a common understanding and viewed as 'true' even with no basis in reality.

This is one of many possible explanations for why humans so often believe in spirits, ghosts, angels, trolls, gods and other beings that lack reality-basis. These are simply cases of agreed-upon knowledge of how the world works; knowledge that happens not to relate to reality.

All these differences between biology and culture have to do with the difference between genes and 'memes'. This may be a fundamental problem when talking about cultural information as transmission of knowledge particles.

Australian philosopher Peter Godfrey-Smith has suggested that a better approach would be to speak about 'Darwinian populations', where this would be 'a collection of causally connected individual things in which there is variation in character, which leads to differences in reproductive output […], and which is inherited to some extent. Inheritance is understood as similarity between parent and offspring, due to the causal role of the parents' (Godfrey-Smith 2009).

Godfrey-Smith's suggestion of 'Darwinian populations' is actually meant to replace the central role of genes in evolutionary theory, but here I think he overstates his case, missing that reproduction and selection is all about genetic transfer of *information* (Dennett 2011). More fitting, perhaps, would be to utilize this approach in cultural evolution (not discounting the importance of information transfer, of course, but using his type of population thinking). However, this is an empirical question for future research.

But there is another difference between cultural and biological evolution that should be noted: how variation is created. In biology, variation comes about exclusively through random processes such as mutation and recombination of the genetic material. In culture, however, another possibility exists besides randomness: design.

Design of cultural elements doesn't need to be *intelligent* design; the examples of failed inventions seem infinite—only the number of attempted perpetual motion machines is enormous. But surely there is room for intelligent and deliberate design; just think of the evolution of the personal computer where it is hard to envision any random input whatsoever (though this may be just due to my poor imagination on this point).

Though the creation of variation (that occurs through random change, recombination and design) and the transmission process (where there is no material information carrier of cultural information that enters directly into brains) are different in cultural evolution, the selection process seems very similar in that cultural traits are selected both according to how well they function and how well they spread.

These two selection factors are the same as those that exist in biology: natural and sexual selection. While natural selection is based on function—how well wings help you fly, talons help you hunt, or legs make you run or hop—sexual selection is about getting a mate. Sexual selection thus produces traits that are either attractive to the opposite sex (peacock tails, bright colors, fancy displays) or dangerous to competitors of the same sex (horns, fangs or superior size). While natural selection is about function, sexual selection is just about getting to mate.

The parallels to cultural selection are obvious. Some cultural traits (boats, plows, elevators, airplanes) have a function that can be better or worse; that can work or be non-functional. Other traits have no such functional filter but are selected on how well they can transmit from one brain to the next. This encompasses musical pieces, art and literature. There are, just as in biology, traits that are selected according to both, such as clothes, where there is both a functional element (warmth) and a transmission element (fashion).

A classic study in this regard was made by Stanford biologists Deborah S. Rogers and Paul R. Ehrlich who analyzed two sets of related cultural traits, functional and symbolic design features for Polynesian canoes, to see if these evolved at different rates. Rogers and Ehrlich could confirm that functional features that are continually tested against the environment were more constrained and thus evolved more slowly than symbolic designs (Rogers and Ehrlich 2008). The same pattern can be found in biology, where functional traits evolve more slowly than weaponry or decorative traits (Seddon et al. 2013).

In culture, however, there are also completely arbitrary traits that are transmitted, such as table manners, or what side of the road you drive on, or what color clothes to wear to a funeral, or what music to listen to be considered 'learned' or 'cool'. These underline the pattern that we don't have to actually understand a trait to carry and transmit it. What difference does it make if you eat with fork and knife, chopsticks, or your hands? Nevertheless, making the wrong choice in the wrong social setting—breaking an established behavioral norm—can be a traumatic experience.

The above-mentioned differences between biological and cultural evolution are large obstacles to the project of transferring analytic methods from biology to culture. It is currently obvious that tools of analysis cannot simply be transferred from biology to culture but have to be developed on their own within the field. This research is currently booming, also evidenced at the place where I work—the Centre for the Study of Cultural Evolution at Stockholm University in Sweden.

It is a bit strange that no coherent and unified theory of cultural evolution has been formulated during the time that human culture has been studied. Instead, the sciences that concern themselves with human expressions have splintered into a number of different subjects. History for things that have already happened, political sciences for how to govern societies, law for judicial matters, anthropology for the study of different cultures, linguistics for the study of languages, and so on. But there exists no real unified and coherent theory of cultural evolution.

However, some patterns can be discerned through simple observation. For example, cultural change goes faster and faster throughout history. Science reveals more and more, ever more books are published, musical styles now replace each other so quickly that some are barely known before they become unknown (Enquist et al. 2008). Why does cultural change accelerate?

Partly, it seems to be due to the number of people. If each person gets one idea, the exponential growth of humanity will in itself result in an exponential growth of cultural traits. The same is true for the storing possibilities of culture. More people can remember more things, together.

But culture is also constructed of existing culture. If you have a spear and a rope it isn't too great a step to harpoon production. Similarly, you can make cream, butter, yoghurt, cheese and a number of other products out of milk, but you first need domesticated cows for milk. Culture breeds more culture.

Finally, it actually seems as we are becoming smarter and smarter. This pattern is termed the 'Flynn effect' after its discoverer, James R. Flynn, a professor of Political Science from New Zealand. He has shown that IQ has increased continuously since the beginning of IQ tests: the bottom level keeps increasing. If you standardize by setting the last measured IQ results as having a mean of 100, then Great Britain in 1942 had an average IQ of 73, Holland 1952 had an average of 70, and Norway 1953 had an average of 88.5 (Flynn 2007).

This increase of average IQ of about three points per decade can be because more culture actually makes you smarter, that schools are getting better and better, or in that the health properties of food keep improving. Several causes together probably lie behind the pattern, but it is an encouraging one. (We have data only from the last 60–80 years, so the results must be interpreted circumspectly.) We have not only developed advanced culture because we are smart; to a degree we have become smart because of our advanced culture.

One question that has intrigued researchers of cultural evolution is to what degree cultural evolution is free from biological evolution. This differs between traits. Neither survival nor reproduction is influenced by whether you speak Swedish in a Swedish-speaking environment of Finnish in a Finnish-speaking environment. Language thus seems 'free' to evolve.

Language change can irritate the conservatively minded. The Swedish archbishop Laurentis Petri, who served during the reign of King Gustav Vasa when the king abolished the Catholic Church and made Sweden a Protestant country, wrote that loan words made 'our Swedish language corrupted and distorted' while 'we have as good words as those imported'.

As is evident from comparing the Old Swedish with current Swedish, he is of course right—Swedish language has changed a lot since his days. But you are hard pressed to find any Swede who suffers because we don't speak the same language as the venerable archbishop. In fact, modern Swedish contains more words of German origin than words derived from Old Swedish—just what Petri warned against. This distinction, to have more imported than retained words, is something Swedish shares with English, which has more words from French than words of ancient English descent (Parkvall 2009). This has not hampered in the least our capability of expressing ourselves.

In some manner, all traits are limited by our biology, even our languages. We cannot, for example, construct a language and speak using smells, or through blinking with lamps in the infrared—we wouldn't be able to see such signals. But it works well to communicate using sign language. So it is a fine balance when discussing exactly how 'free' culture is to evolve. 'Free within limits' is a better phrase, which immediately makes everything much messier.

Different researchers stress different aspects when contemplating this question. American anthropologist John Tooby and his wife, evolutionary psychologist Leda Cosmides, have for example written that 'Human brains, human behaviors, human artifacts and human culture are all biological phenomena' (Tooby and Cosmides 1992). But anyone with language capability can learn Swedish or Kiswahili, regardless of genetic heritage, so how could this be true? Their point is that we wouldn't be able to learn *anything* if our biology wasn't as it is—language is limited and to some degree formed by our biological abilities.

Another researcher, Canadian Anthropologist Joseph Henrich, instead stresses the duality in *dual inheritance theory*, writing that 'The cultural evolution of increasingly sophisticated tools and practices first drove the emergence of this cognitive ability, and then the two entered into a culture-gene coevolutionary duet.' Cultural progress drives the evolution of cultural competencies which in turn drives cultural progress and so on. 'Our bodies have been shaped by fire and cooking, but we have to learn from others how to make fire and cook (Henrich 2016). But this does not explain why Swedish and Kiswahili have evolved into two different languages? Again, without having evolved language capacities we wouldn't be able to use language at all, and surely these capacities co-evolved with language, as did many of our psychological peculiarities as a species, such as or propensity to over-imitate.

Finally, some researchers—almost everyone in the social sciences and humanities—stress that cultural change is a process free from biological constraints. They of course also have a point, in that phenomena like music styles, governance systems, fashions, languages, alphabets, computers and so on, develop (almost) independently of biological constraints. Again, the best description is 'Free within limits' in that music styles, for examples, do not spill over into ultraviolet compositions, or governance systems spill over into alphabets.

An American anthropologist at Stanford University, William Durham, has outlined all potential interactions between cultural and biological evolution. There are five (Durham 1991).

1. *Genetic mediation*—genes give culture. What colors we name is given by the human visual system. All languages have terms for light and dark, and then it is green, red and blue, but never ultra-violet or infrared. This is because our visual receptors react to light (rods) or specific wavelengths (cones) that match the green, red and blue. We lack receptors for ultra-violet or radio waves. First genes, then culture.
2. *Cultural moderation*—culture gives genetics. North Europeans and certain African peoples with a long history of husbandry have a genetic adaptation that allows us to digest milk even as adults. This mutation is about 10,000 years old. An ecological possibility—to raise cattle for their meat—opened up for natural selection where those who could digest milk and thus also could utilize more then just the meat from the cows, got a survival advantage over those who couldn't. Other peoples, for example East Asians, without a similar history, cannot do this. Husbandry emerged rather recently in human populations, so the evolution of lactose tolerance has been relatively quick (Laland and Brown 2002). First culture, then genes.
3. *Amplification*—culture amplifies genetic traits. An example here is the incest taboo that works as a mechanism against inbreeding. A natural aversion towards mating with close kin is amplified by cultural traditions.
4. *Neutrality*—culture is completely free from genetic influence. Examples are what language you speak and what music you prefer, where culture is free to change—within limits set by the cultural trait itself.
5. *Opposition*—culture is negative for the survival and copying of genes. An example is the ritual cannibalism in Fore tribes on Papua New Guinea, where dead relatives were eaten to keep the 'life force' within the tribe. This resulted in an epidemic of the disease *kuru* that is contagious through defective proteins in the brain of those who catch the disease. A deadly cultural tradition.

This last example exemplifies that cultural traits that affect survival and reproduction and that thus affect natural selection—biological selection—can develop in ways that are not all too easy to understand from a biological perspective.

Another example is that the majority of Western societies, now with more resources and better health than medieval kings could dream of, reproduce very little. Some countries have such a low reproductive rate that the population actually decreases. In the end, it may thus be cultural evolution that saves us from overpopulation.

Similarly, it is genetic suicide to opt for celibacy (if you do it right), but still the Catholic priesthood and a number of monk and nunnery orders persist around the globe—most members of these never procreate. Why are people copying this cultural trait? Other people take drugs and participate in lethally dangerous sports, despite the risks.

Possibly, it has to do with status-hunt gone awry. Being a priest in the Catholic Church used to be a highly respected position in society, before all the child-abuse scandals. In this way, 'the idea of celibacy' could spread, despite the lethal effects on genes [of course, everyone didn't live in celibacy even though they pledged to do so; there have even been popes who fathered children (Lawe 2000)]. To challenge death by taking drugs or participating in highly dangerous sports are both 'cool' things to do in certain groups of youth, favoring the spread of these ideas even if they can be detrimental from a genetic perspective. Similarly, the lethal idea of ritual cannibalism was maintained in the Fore tribe because of a strong idea that this was the way life force was retained.

None of these ideas can have evolved through natural selection of genetically influenced behaviors. To understand traits such as these you have to shift your perspective to study exactly what replicators are favored by the spread of these traits. Drugs, celibacy, postponing having children, martyrdom, and ritual cannibalism in no way favor the genes of their bearers. The only way they spread is if the ideas themselves—the replicators, the memes—are favored.

It has even been suggested that agriculture emerged as a self-replicating idea and not at all because it favored the people who adopted it. Examining the archaeological material, agriculture seems not to have been progress at all. Skeletons in the archaeological record become smaller and worn down by the monotonous hard labor of farming, teeth were ground down from chewing bread baked on meal mixed with stone meal from the milling process, and few of the first farmers lived longer than 30 years (Blackmore 1999). It has also been found that the crops of the first farmers cost more in energy input without providing more calories than the hunter-gatherer existence (Bowles 2011).

In an article about the development of agriculture in the scientific journal Nature the American scientist and author Jared Diamond writes this about the first farmers: 'If they had actually foreseen the consequences, they would surely have outlawed the first steps towards domestication, because the archaeological and ethnographic record throughout the world shows that the transition from hunting and gathering to farming eventually resulted in more work, lower adult stature, worse nutritional condition and heavier disease burden' (Diamond 2002).

Why would anyone opt for farming with that future in mind? It takes about 4–6 h for a hunter-gatherer to gather the daily nutrient requirements, far less than the time required for an average farmer (Sahlins 2005).

But with the new replicators—ideas, memes—we don't have to look for a genetic advantage. For the idea of agriculture to spread it is enough if it *seems* as if agriculture is beneficial. And what did the farmer's neighbors see, for example the neighbors of the ancient Egyptians? Pyramids and other grand architectural feats, field after field filled with crops, rich people wallowing in food—the rich are more visible than the poor simply by being rich.

This extra replicator—ideas, memes—is unique in the animal world. Humans have language, an unsurpassed ability to learn from others through a unique ability of thought and better learning mechanisms, memory, judgment ability, planning and creativity than in other animals. These abilities are all young in a biological perspective.

But what is special with humans—what separates us from all other animals—is what these abilities make possible: cultural evolution. It is the cultural evolutionary process—unique to mankind—that is the foundation for the incredible achievements of humans.

Cultural Evolutionary Explanations of Cooperation

In the chapter *Humanity—the paragon of cooperation?* we saw how cultural evolutionary explanations may be the key to understanding the unique cooperation of humans. Scientists researching how genes and culture have evolved together suggest that cultural conformism and punishment of defectors have created a new environment that has resulted in cooperation for the good of the group being written into our genome.

But there is another explanation for the exaggerated aspects of human cooperation, this time disregarding genes.

Remember the factors that best explain the difference between cultures that are generous in the dictator game and those that are not? The dictator game, if you remember, is the game where the first player gets a sum of money and then can choose to share this sum with another, unknown player—or to refrain. Apparently irrationally, some players choose to share their money.

The factors that had explanatory value were how well the society where the experiment was carried out was integrated into the global market economy and if the participating players belonged to a world religion (i.e. Islam and Christianity). Belonging to the market economy as well as to Islam and Christianity are both cultural factors. But not only that, they are united by one common trait: *they all regard cooperation very highly*.

According to Christianity, all who have accepted Jesus as their Lord and savior belong together. As St. Paul phrased it 'There is neither Jew nor Greek, there is neither bond nor free, there is neither male nor female: for ye are all one in Christ

Jesus' (Galatians 3:28). Christianity has unity and belonging as a central part of its teachings—regardless of if you are family or friends, you are 'one in Christ Jesus'. The same is true for Islam, where the worldwide community of Muslims even has a name: *umma*. This is not regarded only as a community based on religious faith and life style, but also as a political community.

The devotion to God or a sacred cause may serve as a signal of group identity and may result in irrational intractability that is good for the group over the long term. Otherwise mundane conflicts may acquire a status of being sacred, making them immune to material incentives. Thus, religion may have enabled large-scale cooperation (and conflict) through promoting prosocial behaviors (Atran and Ginges 2013).

It has even been suggested that ideas of 'Big Gods' supervising us was a necessary step during our cultural evolution, when going from small tribal bands to large-scale societies. This because the thought of a super-natural overseer made people behave even when they were not directly monitored by family and friends (Norenzayan 2013).

Concerning a market economy, it *requires* cooperation to be able to function. Meeting and interacting socially are necessary for trade, but you also have to have an agreement over the rules of how this happens. You have to give and take in negotiations, evaluate and haggle, accept or reject proposals.

The interesting thing is not so much that these teachings advocate cooperation. The central issue is that cooperation is an *integral part of the teachings*. Time, then, to shift perspective to the cultural traits (call them 'memes' or whatever you want). If you were a selfish cultural trait, a selfish idea, what would be the best way to spread through a population—the best way to be contagious?

To spread, the individuals carrying the cultural trait must first and foremost be made to interact with other people. There is a marked advantage to being part of a larger cultural unit wherein is included the idea of regular meetings, such as Friday prayers in a mosque, Sunday service in a church, or market days.

It is also advantageous to be part of a cultural bundle wherein cooperation is favored. For example, favorable treatment of people of the same faith or fellow merchants, effective propaganda for the virtues of the faith or system, and a striving to conform to the rules. Punishment of defectors also plays a central part, either through threats of hell and damnation, excommunication, beheading, or—in the case of the market economy—loss of trust and trading possibilities, or judicial consequences.

Note that ideas such as religions and the market economy can be subdivided into smaller collaborative component ideas. Meme theorists call such linked ideas memeplexes, or cultural systems.

There are many other cultural systems than Christianity, Islam and the market economy that have cooperation as an integral part. Examples are political ideologies, citizenships in countries, or membership of regiments.

All these ways of belonging constitute cooperation in a way that makes us feel a part of 'something more' and evokes feelings to show and celebrate belonging. What is it really that evokes a feeling of euphoria when 'our' country's team scores? Is that a reasonable feeling? It can simply be that the cultural traits that evoke feelings of belonging are the ones that spread best.

Clearly, all cultural systems do not include cooperation. The idea 'live in a house' or 'make stone axe' gives immediate material benefits and can thus on some level be understood biologically. Thus, it is no surprise that nests and tools exist for other animals in nature. However, ideas of how to make houses or stone axes even better spread more easily in humans since we can learn from each other and evaluate the outcomes of different designs. In this manner, functional ideas can be evaluated directly in the interaction with the environment on how well they work.

The ideas 'Bach's Toccata and Fugue in D-minor' or 'Ramones' Rockaway Beach' on the other hand provide immediate emotional responses in people who love those kinds of music; likewise with fancy clothes. People who get feelings they enjoy from music and clothes will adopt these ideas for hedonistic reasons—because they enjoy them. Exactly *what* music or *what* clothes are favored is not completely random—there are fashions. And fashions are contagious, possibly because we like to copy traits from people we see as successful.

The key insight is that cultural traits spread as if it is they and not humans who are the cooperating units. This is cooperation in exactly the same way as genes cooperate. Some traits (genes and memes) spread better than others because they work better and spread better together. In some cases, these traits (genes and memes) work only in the presence of other traits.

Note that it isn't the case that cultural traits actually *are* selfish and willfully run our lives. But it helps our understanding if we can look at these traits *as if* they were selfish, *as if* they ran our lives as if we were mindless robots. Agency is implied—not present in reality.

The thought of selfish cultural elements can be unpleasant. Philosopher Peter Godfrey-Smith calls this type of thoughts 'Darwinian paranoia'—beliefs that we are governed by agents out of our control, such as memes (or genes, for that matter) (Godfrey-Smith 2009). As the philosopher Daniel Dennett has put it: 'I don't know about you, but I am not initially attracted by the idea of my brain as a sort of dung heap in which the larvae of other people's ideas renew themselves, before sending out copies of themselves in an informational Diaspora. It does seem to rob my mind of its importance as both author and critic. Who is in charge, according to this vision—we or our memes?' (cited in Gleick 2011).

But not being in charge is not just a manner of speaking or interesting thought experiment. There are many signs that humans have gone through a sort of 'self-domestication' process during our evolution. Ordinary domestication is when natural selection goes from being mainly about adapting to the environment to mainly adapting to social cues. This is what has happened to dogs, cats, sheep, cows, and so on—they have been selected to function well with humans. Many traits evolve predictably from such selection; traits like docility and tameness, reductions in tooth size, prolongation of juvenile behavior, increased playfulness, lower stress and scare response and reduction in brain size—all making domesticated species less easily scared and less aggressive.

These are all changes that have occurred during human evolution. But hold on, what's that about brain size *reduction*—didn't we previously go through evidence of a brain size *increase* in the human lineage the last two million years? This is still true, but interestingly, there is an actual observable brain size *reduction* in the human lineage during the last 200,000 years—exactly the period during which there is clear evidence of cultural evolution. During this time, we lost brain volume amounting to about the size of a tennis ball (Hood 2014; Trut et al. 2009).

Daniel Dennett has proposed that what has domesticated humans is culture. When it comes to genes, what is good for an individual is good for the genes (though the contrary is not necessarily true, as we have seen). The situation is not at all the same for 'memes'—the fate of their copying success has only little necessary connection to the fate of their bearers. Think, for example, again of celibacy or martyrdom. Even though these traits are 'genetically deadly' traits in their correctly executed form in the same manner as a sterilizing or deadly disease, they are maintained and spread in many populations, if the carriers are viewed as role models, even if they are biologically harmful.

If you think about it, most of us spend our lives not maximizing our biological success through getting as many children as possible, but working in the interest of 'memes', by serving democracy, the market economy, freedom, equality, knowledge, or just trying to get a comfortable life. There has been a shift away from biological interests to cultural.

The competition of ideas for a place in our brains is natural selection, but on cultural traits—*cultural selection*. Cultural traits that provide emotional satisfaction or just work better have a greater chance of spreading than cultural traits that provide emotional discomfort or don't work. Are we even the masters of our own fates anymore, or are we servants of the memes?

Traits that favor cooperation and cultural systems that contain cooperation can be yet another explanation for why humans cooperate so much, explaining why the cooperative impulse differs so much between cultures—cultural *norms* about cooperation differ.

Networks

But if people just are information carriers and soullessly act as containers for collective knowledge, then human interactions should be analyzable as if humans were just soulless nodes in social networks. How do ideas spread across such networks? Why, for example, are people so overweight these days? Some say that there is an epidemic of obesity in the USA. But being fat isn't contagious. How can it be an epidemic?

Research on human networks indicates that we are indeed influenced by our friends. This may not be surprising, but in extension it has some startling consequences. Our friends and their friends transmit ideas, behaviors and views that influence us and that we in turn transmit further. Researchers studying networks have shown that there is a much larger risk for you to become overweight if you have a friend who has become overweight. So yes, being overweight is contagious. Or rather, behaviors and ideas that lead to obesity are contagious.

And it doesn't even have to be your own friends that cause the problem. You actually run a greater risk of becoming overweight if you have a friend of a friend who has become overweight. The same is true for eating disorders, divorce, suicide, pregnancy, voting, sick leave, unemployment, unselfishness and—happiness. It has a measurably larger effect on your happiness if your friend becomes happier than if you win $5000 on the lottery (Fowler and Christakis 2011). Luckily, happiness is more contagious than depression (Nowak and Highfield 2011). [There has been some debate about this research, but social contagion has lately gained support through further analyses (Kramer et al. 2014).]

One especially interesting observation is that the inclination to cooperate also is contagious. In the chapter on cooperation games we saw that human exaggerated cooperation can be maintained through punishing cheaters who invest too little. Two American researchers, James Fowler and Nicholas Christakis, have found that the proclivity to invest a lot in cooperative games is 'contagious' in networks. Players who have just had an interaction where another player invested a lot tend also to invest a lot in the next round. This effect was measurable in three steps—there is a noticeable difference if the player has played with someone who has played with someone who invested a lot (Fowler and Christakis 2011).

Martin Nowak, the researcher trying to dethrone Hamilton's kin selection theory, has even shown, as we saw earlier, that there is a simple mathematical rule for the spread of cooperation in networks. This happens when the ratio between benefit and cost (*b*/*c*) when cooperating is larger than the average number of friends (neighbors) in the network. This is called network reciprocity, or spatial selection (Nowak and Highfield 2011)—it is easier to cooperate if you form groups of cooperating individuals. This equation has received empirical support in game-theoretical experiments (Rand et al. 2014).

In an especially clever analysis, three German researchers have let people playing multi-player games punish defectors by removing them from their list of partners. The effect was that clusters of cooperating individuals formed in the larger network structure. Cooperation *in itself* thus became a signal of your trustworthiness—a Green Beard—and this in turn had repercussions on other players' willingness to cooperate with the shunned player (Fehl et al. 2011).

Game-theoretical experiments in the Hadza, hunter-gatherers in Tanzania, have further shown that they have social connections mainly with individuals giving about the same amount as they themselves do (Apicella et al. 2012). What is cause and what is effect is, however, hard to disentangle—do members of the Hadza choose to interact mainly with people who are equally cooperative or is cooperation contagious?

How networks between people look in reality is not only determined by how cooperative someone is, but also by how social someone is. Some have many friends and acquaintances while others have few. Some introduce friends and family to others, becoming embedded in a large interconnected network while others keep their circles separate and in this way may serve as bridges between different social networks. Some are in the center of their networks, others are peripheral.

But most of us like to be together with people like ourselves—you can even see this in the observation that people are more genetically alike their friends in comparison to the population in large (Fowler et al. 2011). And most of us more easily adopt traits, ideas and thoughts from our friends and family than from total strangers. The structure of social networks turns out to have great influence over how easily disease and ideas spread. Is the network loosely or strongly interconnected, is the source individual located in the center or the periphery? Such differences have a tremendous impact.

Knowledge of how ideas and disease spread in networks can be utilized to optimize such undertakings as vaccination campaigns. To inhibit measles it is normally estimated that about 90% of the population has to be vaccinated. But there is a cheaper, more creative alternative.

Ask a random selection of people to name one friend. Then vaccinate these friends. Why would this work? Because people who have many friends tend to be named as friends more often than people with few friends (yes, your friends have more friends than you do). If you vaccinate these more commonly named friends until 30% of the population has been vaccinated, calculations show that you achieve the same protection against measles as if you randomly chose 99% of the population (Christakis and Fowler 2009).

The evidence supporting a role of spatial selection for the evolution of cooperation between humans is mixed, however. Further work is needed to clarify exactly when and in what way spatial selection promotes cooperation (Rand and Nowak 2013).

Investigating cooperation between individuals according to the network model turns human societies into a structure very similar to a gigantic brain where individuals are neurons and friendships are synapses (the contacts between neurons). When an idea, trait or disease emerges in some individual it spreads between nodes similarly to how signals are transmitted across neuronal networks. Social networks in this way can be said to achieve a kind of collective 'intelligence'. (On the other hand, brains function a bit like networks of ant paths, where paths used more frequently are strengthened and those used more seldom weakened—analogies can be used to support almost any argument.)

Networks can also have their own memories—information not immediately available to any single individual. Examples of this are common traditions such as rituals; no one knows all traditions but everyone knows a bit; in this way the network can know them all.

Societies can therefore make decisions through combining many individuals' acts. This is how election results and market values of products are determined. Social networks have a 'memory' of their structure and remain intact even if individual persons disappear. They maintain knowledge relatively independently of the separate parts.

From these observations some extrapolate that human societies can be viewed as 'super organisms' with their own characteristics, memories, goals and life histories. Such a human super organism can retain a gigantic amount of information.

Again, this is not about adding similar units and thus ending up with a larger collection of units, but about individual people adopting different specialized competences that all contribute to collective knowledge. Specialists form coalitions with each other and together with other coalitions of specialists form an enormously complex web of interactions: society.

With language as the means for communication, human societies work as composite, 'thinking' units—collective brains. Just as no single nerve cell needs to understand the object of a thought for thought processes to work, no single human needs to understand all parts of an invention for it to work. Ever-increasing specialization is probably one of the most important processes in the development of modern societies and explains the explosive evolution of these.

The British author Matt Ridley has exemplified this process with the computer mouse—who was the inventor of that? (Ridley 2010a, b) As was the case with the pencil and the breakfast sandwich described in an earlier chapter, no one needs to understand all processes required to make a computer mouse. A computer mouse is instead a composite of many inventions that together resulted in a new invention (separate inventions such as plastic, steel, electronics, hardware, software). Ridley describes this phenomenon as 'ideas having sex'. Together, we humans can, through similar combinations of skills, solve problems and create things that no human can understand by herself—innovation depends more on teamwork than on individual intellectual abilities.

If you regard individuals as nodes in a network it seems as if we don't choose our ideas ourselves. You can't choose your body and similarly you can't choose your culture—it is something you are born into. Ideas are contagious between humans—there is no central planning, no governing complex that plans what humanity can come up with next. Just as many nerve cells together result in a brain and a 'self', we now form a collective 'us' that can work miracles—and disasters.

As Richard Dawkins expresses this notion in *The Selfish Gene*, we are vehicles for our genes, machines that are programmed to transmit the program ever onwards. But we are more than that, because we are also nodes in a network that stores and spreads information, nodes that are continuously reprogrammed by contagious ideas. The tools that these sometimes conflicting genetic and cultural programs use to steer us are our preferences and our wishes, our views and our will, our thoughts and our emotions.

The Software

Returning to individuals, we have now, by including common cultural elements—ideas, memes—in our collection of examples of cooperation gotten to the last component of the building blocks of what constitutes a human. To provide some analogies, we now have the music that can be played on the cello, the motif of the painting, the content of the text, the software of the computer—the 'you' in your 'self'.

The computer analogy is especially tempting because of the obvious likeness between the two information processing units: the brain and the computer. Nevertheless, it is just an analogy, this has to be remembered. With language and culture our own information processing units can be reprogrammed. Language can be viewed as the operating system of the brain, the interface through which ideas can be transmitted from brain to brain. This interface is the first thing that has to be learnt before you can receive further information—language learning is the brain's bootstrapping process for cultural competence.

There is no fixed border between language and culture, but neither is there are fixed border between operating system and computer program. In principle, culture can be viewed as the ideas that are transferred, the information in itself. How well we can adopt this new information depends on what we have learnt earlier—that is why schooling is so important.

As anthropologist Joseph Henrich phrases it, 'You, for example, probably received a massive cultural download while growing up that included a convenient base-10 counting system, handy Arabic numerals for easy representation, a vocabulary of at least 60,000 words (if you are a native English speaker), and working examples of the concepts surrounding pulleys, springs, screws, bows, wheels, levers, and adhesives' (Henrich 2016).

The information we incorporate rebuilds our brains, totally outside our own conscious control, something that both enables and limits what we can learn next. We can also give others and ourselves instructions and in this way deliberately store information and processes.

But that we have some control of what we learn doesn't mean that we have direct control over the hardware. We can't get our neurons to turn off and on more quickly or slowly, or individually. Actually, we have no control whatsoever over our hardware. As the American physicist and cognition researcher Douglas Hofstadter has expressed it: 'We cannot make ourselves smarter by an act of will; we cannot learn a new language as fast as we want; we cannot make ourselves think faster than we do; we cannot make ourselves think about several things at once; and so on' (Hofstadter 1979). We may feel we have control of our own thoughts, but this is true only within fixed limits.

There exist physical manifestations of culture—culture exists both in the brain (language) and outside (violin concertos)—but the special thing with for example the fourth movement in Berlioz's *Symphonie fantastique* isn't exactly what the instruments play and when, but the full experience, that when understood transfer us to walk beside a man sentenced to death on his way to the guillotine. Or we just hear a melody—all depending on our prior knowledge, or lack thereof. As Goethe put it in one of his maxims (106): 'What you don't understand, you don't possess' (von Goethe 1833, Maxim 106).

But even non-material ideas have a sort of physical manifestation. Originally, no other medium existed than the brain for ideas to attach themselves to. What happens when we learn things is that nerve cells fire or don't when we are processing information with term memory and thought processes; connections are strengthened or weakened and genes turn on and off in different nerve cells when long-term memory becomes established. These are physical manifestations of cultural traits.

Thus, it is fitting to liken ideas, thoughts and memes to viruses. Like viruses, memes cannot reproduce on their own, and like viruses, cultural traits use their host organism's genes to reproduce. Like viruses, cultural traits are not actually alive but then again not exactly dead. Like viruses, ideas evolve separately from the organisms that become infected.

The British psychologist Nicholas Humphrey expresses it thus: 'When you plant a fertile meme in my mind you literally parasitize my brain, turning it into a vehicle for the meme's propagation in just the way a virus may parasitize the genetic mechanism of a host cell. And this isn't just a way of talking—the meme for, say, 'belief in life after death' is actually realized physically, millions of times over, as a structure in the nervous system of individual men the world over' (Gleick 2011).

The characteristics that ideas result in, do not, however, always have an external physical manifestation in the way viruses have. What is the physical manifestation of how a symphony sounds, or the thought of an immortal soul? This can make culture difficult to understand.

No cultural traits exist in isolation but always in relation to already existing information in both individual brains and societies in large. Information transfer is context-dependent. Thus by necessity some cultural information is inaccessible to us—there are so many cultures, so many memes—and too little time to learn to understand them.

But *too much* knowledge can also stand in the way. I still remember the force of the experience when I slowly grasped the explanatory power of evolutionary theory as listening to lectures and reading books kept entering my experience. That sense of wonder and revelatory understanding are unattainable for my current self. The experience of an intellectual awakening has been experienced and the possibility to re-live it simultaneously disappeared.

How can someone with even a minimum of scientific knowledge now appreciate the explanatory power that once existed in antique myths and creation tales? We know more now and with that knowledge, culture that was crucial, revelatory and of immense importance to previous generations disappears.

Culture changes according to what seems to be an evolutionary process. But the evolutionary aspect of the process means that it leaves a trace of ideas, memes, thoughts, feelings and other phenomena that didn't survive testing, the passage of time and the meeting with reality.

Some things presented in this book, including some scientific results and hypotheses, will also end up on history's trash heap of failed ideas. Probably even some of the explanations of mankind's phenomenal cooperation. Which ideas? We don't know, but many scientists around the world are working on it. Knowledge progresses; as the trash heap of discarded ideas grows, so does our knowledge.

References

Apicella, C. L., Marlowe, F. W., Fowler, J. H., & Christakis, N. A. (2012). Social networks and cooperation in hunter-gatherers. *Nature, 481*, 497–501.

Armus, H. L., Montgomery, A. R., & Jellison, J. L. (2006). Discrimination learning in *Paramecia (P. caudatum). The Psychological Record, 56*, 489–498.

Atran, S., & Ginges, J. (2013). Religious and sacred imperatives in human conflict. *Science, 336*, 855–857.

Basalla, G. (1988). *The evolution of technology*. Cambridge: Cambridge University Press.

Blackmore, S. (1999). *The meme machine*. Oxford: Oxford University Press.

Bonner, J. T. (1980). *The evolution of culture in animals*. Princeton: Princeton University Press.

Bowles, S. (2011). Cultivation of cereals by the first farmers was not more productive than foraging. *Proceedings of the National Academy of Sciences, 108*, 4760–4765.

Boyd, R., & Richerson, P. J. (1985). *Culture and the evolutionary process*. Chicago: University of Chicago Press.

Boyd, R., & Silk, J. B. (2009). *How humans evolved* (5th ed.). New York: WD Norton & Co.

Carneiro, R. L. (2003). *Evolutionism in cultural anthropology: A critical history*. Oxford: Westview Press.

Cavalli-Sforza, L. L., & Feldman, M. (1981). *Cultural transmission and evolution*. Princeton: Princeton University Press.

Christakis, N. A., & Fowler, J. H. (2009).*Connected: The surprising power of our social networks and how they shape our lives*. London: Little, Brown and Company.

Dawkins, R. (1976). *The selfish gene*. Oxford: Oxford University Press.

Dawkins, R. (1989). *The extended phenotype*. Oxford: Oxford University Press.

Deacon, T. W. (1997). *The symbolic species: The co-evolution of language and the human brain*. Allen Lane: The Penguin Press.

Dennett, D. C. (2003). The Baldwin effect, a crane, not a skyhook. In B. H. Weber & D. J. Depew (Eds.), *Evolution and learning: The Baldwin effect reconsidered* (pp. 69–106). Cambridge, MA: MIT Press.

Dennett, D. C. (2011). Homunculi rule. *Biology and Philosophy, 26*, 475–488.

Diamond, J. (1997). *Guns, germs and steel: A short history of everybody for the last 13,000 years*. London: Vintage.

Diamond, J. (2002). Evolution, consequences and future of plant and animal domestication. *Nature, 418*, 700–707.

Durham, W. H. (1991). *Coevolution: Genes, culture, and human diversity*. Stanford University Press.

Enquist, M., Ghirlanda, S., Jarrick, A., & Wachtmeister, C.-A. (2008). Why does human culture increase exponentially? *Theoretical Population Biology, 74*, 46–55.

Fehl, K., van der Post, D. J., & Semmann, D. (2011). Co-evolution of behaviour and social network structure promotes human cooperation. *Ecology Letters, 14*, 546–551.

Flynn, J. R. (2007). *What is intelligence?: Beyond the Flynn effect*. Cambridge: Cambridge University Press.

Fowler, J. K., & Christakis, N. A. (2011). Cooperative behavior cascades in human social networks. *Proceedings of the National Academy of Sciences, 107*, 5334–5338.

Fowler, J. H., Settle, J. E., & Christakis, N. A. (2011). Correlated genotypes in friendship networks. *Proceedings of the National Academy of Sciences, 108*, 1993–1997.

Gleick, J. (2011). *The information: A history, a theory, a flood*. London: Harper Collins.

Godfrey-Smith, P. (2009) *Darwinian populations and natural selection*. Oxford University Press.

Henrich, J. (2016). *The secret of our success: How culture is driving human evolution, domesticating our species, and making us smarter*. Princeton: Princeton University Press.

Hofstadter, D. R. (1979). *Gödel, Escher, Bach: An eternal golden braid*. London: Penguin.

Hood, B. (2014). *The domesticated brain*. London: Pelican Books.

Kline, M. A., & Boyd, R. (2010). Population size predicts technological complexity in Oceania. *Proceedings of the Royal Society, Series B: Biological Sciences, 277*, 2559–2564.
Kramer, A. D. I., Guillory, J. E., & Hancock, J. T. (2014). Experimental evidence of massive-scale emotional contagion through social networks. *Proceedings of the National Academy of Sciences, 111*, 8788–8790.
Laland, K. N., & Brown, G. R. (2002). *Sense & Nonsense*. Oxford: Oxford University Press.
Lawe, K. (2000). Påven som hade nio barn. Forskning & Framsteg 3. http://fof.se/tidning/2000/3/paven-som-hade-nio-barn
Lewens, T. (2013) Cultural Evolution.In E. N. Zalta (Ed.), *The Stanford Encyclopedia of Philosophy* (Spring 2013 Edition). http://plato.stanford.edu/archives/spr2013/entries/evolution-cultural
Lind, J., & Lindenfors, P. (2010). The number of cultural traits is correlated with female group size but not with male group size in chimpanzee communities. *PLoS ONE, 5*, e9241.
Lind, J., Lindenfors, P., Ghirlanda, S., Lidén, K., & Enquist, M. (2013). Dating human cultural capacity using phylogenetic principles. *Scientific Reports, 3*, 1785.
Myrdal, J. (2004–2005). Motsatstänkandet I praktiken. Ett historiefilosofiskt försök. *Folkets Historia*, 23–33.
Norenzayan, A. (2013). *Big gods: How religion transformed cooperation and conflict*. Princeton: Princeton University Press.
Nowak, M. A., & Highfield, R. (2011). *Supercooperators*. London: Canon Gate.
Parkvall, M. (2009). *Lagom finns bara i Sverige och andra myter om språk*. Stockholm: Telegram Bokförlag AB.
Proffitt, T., Luncz, L. V., Falótico, T., Ottoni, E. B., de la Torre, I., & Haslam, M. (2016). Wild monkeys flake stone tools. *Nature*.
Rand, D. G., & Nowak, M. A. (2013). Human cooperation. *Trends in Cognitive Sciences*, 413–425.
Rand, D. G., Nowak, M. A., Fowler, J. H., & Christakis, N. A. (2014). Static network structure can stabilize human cooperation. *Proceedings of the National Academy of Sciences, 111*, 17093–17098.
Rendell, L., Boyd, R., Cownden, D., Enquist, M., Eriksson, K., Feldman, M. W., et al. (2010). Why copy others? Insights from the social learning strategies tournament. *Science, 328*, 208–213.
Richerson, P. J., Boyd, R. T., & Henrich, J. (2003). Cultural evolution of human cooperation. In P. Hammerstein (Ed.), *The genetic and cultural evolution of cooperation* (pp. 357–388). Cambridge: MIT Press.
Ridley, M. (1996). *The origins of virtue*. London: Penguin.
Ridley, M. (2010a). *The rational optimist: How prosperity evolves*. London: Fourth Estate.
Ridley, M. (2010b). When ideas have sex. http://www.ted.com/talks/matt_ridley_when_ideas_have_sex.html
Rogers, D. S., & Ehrlich, P. R. (2008). Natural selection and cultural rates of change. *Proceedings of the National Academy of Sciences, 105*, 3416–3420.
Sahlins, M. (2005). *The original affluent society*. http://www.eco-action.org/dt/affluent.html
Seddon, N., Botero, C. A., Tobias, J. A., Dunn, P. O., MacGregor, H. E. A., Rubenstein, D. R., et al. (2013). Sexual selection accelerates signal evolution during speciation in birds. *Proceedings of the Royal Society, Series B: Biological Sciences,* 280.
South Tyrol Museum of Archaeology web-site. (2016). http://www.iceman.it/en/
Struhsaker, T. T., Cooney, D. O., & Siex, K. S. (1997). Charcoal consumption by Zanzibar red colobus monkeys: its function and its ecological and demographic consequences. *International Journal of Primatology, 18*, 61–72.
Tooby, J., & Cosmides, L. (1992). The psychological foundations of culture. In J. H. Barkow, L. Cosmides, & J. Tooby (Eds.), *The adapted mind: Evolutionary psychology and the generation of culture* (pp. 19–136). Oxford: Oxford University Press.
Trut, L., Oskina, I., & Kharlamova, A. (2009). Animal evolution during domestication: The domesticated fox as a model. *BioEssays, 31*, 349–360.

von Goethe, J. W. (1833). *Maximen und Reflexionen*. Anaconda.
Whiten, A., Goodall, J., McGrew, W. C., Nishida, T., Reynolds, V., Sugiyama, Y., et al. (1999). Cultures in chimpanzees. *Nature, 399*, 682–685.
Wills, C. (1993). *The runaway brain: The evolution of human uniqueness*. London: Harper Collins.
Wilson, E. O., & Lumsden, C. J. (1981). *Genes, mind and culture: The coevolutionary process*. Harvard: Harvard University Press.
Wittgenstein, L. (1958). *The blue and brown books: Preliminary studies for the philosophical investigations*. Wiley-Blackwell.

Epilogue: The Human Super Organism

Abstract This chapter summarizes the book and points out that cooperation is not something beneficial in itself, but needs to be understood in order for its negative aspects to be harnessed. The central idea is stressed, that all cooperation benefits the copying of the replicators—genes or memes.

> As man advances in civilization, and small tribes are united into larger communities, the simplest reason would tell each individual that he ought to extend his social instincts and sympathies to all members of the same nation, though personally unknown to him. This point being once reached, there is only an artificial barrier to prevent his sympathies extending to the men of all nations and races.
>
> Charles Darwin: The Descent of Man (1871)

Every one of us, in both body and mind, is made up of conglomerate upon conglomerate of cooperating units, from the inside out, layer upon layer.

Genes are the base. Life's most basic building blocks contain not only the bauplan of every organism, but also life's most fundamental instruction: COPY ME. From this initial molecular directive evolved the diversity of life. Cooperation between genes, directing such self-replication, is not only basic to life; it can reasonably be posited that this cooperation for self-replication is the best definition of life we possess.

Without this cooperation for self-replication, evolution would not be possible, metabolism and energy circulation would not be needed, and the ability to register ambient conditions would be unnecessary. Self-replication creates order from disorder by repeating the same instruction over and over, driven by the continuous absorption of order from the environment (in the shape of sunlight or other life). Cooperation is thus one of life's most fundamental processes.

Better cooperation between genes means better copying of these genes. Instructions that are advantageous in one environment (e.g. sharp fangs) do not work well in an organism with other life conditions (e.g. herbivory). How well genes succeed in being copied therefore largely depends on how compatible the genes' products are—on how well genes interact with other genes.

Changes in any feature are tested continually by natural selection. Success is measured by the degree in which genes conveying instructions for the changes

P. Lindenfors, *For Whose Benefit?*, DOI 10.1007/978-3-319-50874-0_10

succeed in copying themselves. Genes are the basic self-copying unit—the fundamental *replicator*—whose relative success in spreading copies of themselves is the mark of success for every new feature.

However, the smallest individual building blocks of a human being that can exist independently—at least in a nutrient solution—are the cells. Each cell contains organelles (the 'organs' of bacteria) evolved from bacteria incorporated into our cells that now function as energy factories: the mitochondria. These have their own genome that cooperates with the genome of the cell, to the mutual advantage of both parties. Every cell in our body thus basically consists of cooperating cells within cells that over millions of years have evolved into novel units.

Our bodies' cells, in turn, have conflated in large communities that, like a three-dimensional mosaic, are what constitutes our physical being, where the fate of the individual cells are of secondary importance (they have been 'de-darwinized'; Godfrey-Smith 2009), thile the fate of the replicators (genes) is tied to the success of the vehicles (Dawkins 1976). In the body, every cell knows where it is. Different genes are turned on and off so that specialized cells together can build organs in the correct places. These specialized organs each address a specific task and together assemble a complete human body.

On every level—genes, organelles, cells, organs—the specialization from a common pattern has resulted in entities that have become greater than the sum of their parts. On every level there are strict control mechanisms to handle elements that break cooperation. On every level there exists a surfeit of parts—a surplus of cooperating units—that make the whole more stable than the parts. On every level, the effectiveness of cooperation is measurable by the effect it has on the genes—the replicators. Each level is characterized by a new level of order.

Intricate cooperation occurs in the brain as well. First, millions of nerve cells grow and send out tentacles to contact millions of other nerve cells according to a development plan that is only partly genetically determined. At subsequent stages, some contacts between nerve cells are discarded while others are reinforced.

In this way, the brain structures itself physically. Thereafter, cultural attributes (ideas, thoughts, memes) move in and settle in with other cultural attributes. Ideas and thoughts that cooperate can co-exist, others are forced out. Unfulfilled prophecies about the end of the world are an example of expectations that collide with reality. Traits like that are remodeled or have to exit. Ideas and thoughts also have to function together.

Ideas—socially transmittable elements, thoughts, memes—were the second fundamental replicator after genes to emerge on earth. Ideas can be communicated from individual to individual through imitation and our unique human capability, language. Such ideas and thoughts have produced their own evolutionary process—*cultural evolution*—that takes place in every human brain, but mainly in our collective consciousness. This cultural evolutionary process is unique to mankind and is the explanation for our remarkable success in spreading across all the earth's ecosystems.

Cooperation between individual people exists because it serves either the copying of genes or memes, or both. This is the common answer to the question in the title: For whose benefit?

We are biological beings; we are all animals. Like other social animals, we cooperate with each other according to patterns that benefit the copying of our genes. The simplest form of cooperation between individuals is to co-exist in groups, something that may help us as individuals and as a result simultaneously be of advantage to our genes. This type of cooperation does not imply a contradiction between the interests of the individual and the 'interests' of our genes and is therefore not especially hard to understand. Cooperation exists because it is better that way for all units involved.

Harder to explain is cooperation that involves self-sacrifice in individual organisms. Cooperation of this kind can occur when the self-sacrificing individual favors its own cooperative genes, though in other individuals. This can happen when individuals choose to cooperate with related individuals—relations—where the probability is high that the individual being helped carries the same gene for cooperation as the individual making the sacrifice.

All other cooperation between animals besides kin selection needs to benefit the acting individual directly or the genes that it is carrying directly; *only* kin selection works through genes coding for self-sacrificial behavior *helping copies of themselves in other individuals*.

Genetically influenced cooperation can also occur among unrelated individuals through the repeated exchange of favors between individuals who take turns in making sacrifices, through remembering with whom it has been productive to cooperate. This is the biological explanation of friendship and trust, a kind of cooperation found only rarely in nature but which at least is reported to occur among apes other than ourselves.

However, these three explanations of cooperation—immediate benefit, kin selection and reciprocal altruism—are insufficient to explain all cooperation between humans. People regularly cooperate with unrelated, unfamiliar individuals who they do not expect to meet again. This form of cooperation appears not to exist among animals other than humans.

A potential genetic explanation for this cooperation is termed *indirect reciprocity*. Apologists for this theory say that because of our language ability we have the ability to gossip and therefore can inform others about who are good cooperative partners and who are not. This allows us to keep check on a larger number of people and even gather information about people we have never met.

Another possible genetic explanation is termed *cultural group selection*, through a mechanism called altruistic punishment, where people are punished for freeloading if they do not live up to the cultural norm of exaggerated cooperation that is expected of them. According to this theory, the cultural norms that are maintained over lengthy periods and the punishment of refusants have enabled group selection of genes for 'extra' cooperation between people who do not know each other but belong to the same group. This extreme cooperation in humans may even have co-evolved with our proclivity for waging war.

The inclination for cooperation, however, differs between cultures, indicating that cultural differences also are important to an understanding of the phenomenon. Ideas and thoughts can be transmitted between people through language or imitation. Language may have evolved as a cooperation enhancer, where recognition of speech and specific languages may have functioned as 'green beards'—a mechanism to identify others with the same cooperation-reinforcing property—that helped spread the function. Simply put, it is possible to recognize other language users through the use of language.

While language capability is a cooperation-reinforcing agent, it is also the prime example of an idea that cooperates between (and over the heads of) individuals, regardless of whether we are engaged in cooperation or conflict. It is indifferent to the language you are using whether you are cooperating through trade or you are having a conflict by exchanging insults—language is used in both cases.

When language has been established, culture can be transmitted as vocal instruction rather than imitation of behavior or transmission of complete products. This 'protects' instructions from error in the same way that inheritable genes are protected from the errors that arise in mature cells or the bodies whose recipes they carry. You can concisely inform someone how to grow food, avoiding years of misdirection through your own feeble efforts.

Language unlocks cooperation between other ideas and thoughts, regardless of whether these ideas work in our genetic interest or not. An example of such genetically detrimental cultural cooperation is the Catholic Church's idea of celibacy; despite the almost genetically fatal effect of this lifestyle (for those who actually follow the dictum) it has been retained by the church for 2000 years.

So the question perhaps ought not to be why people cooperate so much; a better question might be to what degree are we being guided by our two replicators: genes and memes?

Love of children and family as well as the strong feelings implied in close friendship and trust between familiar people are consequences of the interests of our genes. In 'The Selfish Gene' Richard Dawkins suggests that individual organisms be visualized as vehicles steered by genes. But he also coined the world 'meme' and discusses the tendency of ideas to spread between brains and be filtered in a kind of evolutionary process.

This second replicator implies that humans are steered not only by genes but also by memes. Is the explanation for extraordinary human cooperation to be found in cooperating ideas?

The idea of cooperation is the most powerful replicator imaginable—a reciprocal replicator that benefits itself to the advantage of all entities it is associated with. Every nexus of ideas that includes the idea of cooperation will do better than similar conglomerates that do not include the concept of cooperation. The cultural qualities that make us behave as their subordinates disperse best.

Human societies can be described as collections of people representing nodes in networks where ideas spread like contagion. These human networks build structures that mostly resemble a kind of brain, or super-organism, with its own memory, its own behavior, interests and idiosyncrasies.

Man is suffused by cooperation—in our physical and mental beings, our families and friendships, our communication and culture, our societies and our thoughts.

Cooperation is our heritage.

Characteristics of Synergistic Cooperation

Some characteristics are common to all instances of cooperation that have resulted in synergy effects—those instances where cooperation has resulted in something emergent, something larger than the sum of its parts; the cooperation between genes, organelles, cells, nerve cells, memes, and between humans in the large collections of individuals we call societies.

American biologist David Queller has made a useful distinction between 'fraternal' and 'egalitarian' cooperation. Egalitarian cooperation is between unlike components, where two different kinds of entities come together bringing different capacities that work well in combination (our bacteria and ourselves, for example). Fraternal cooperation is between like components where like kinds of entities work together for mutual benefit (our body-cells, for example) (Queller 1997). Differences exist between these two types of cooperation, as indicated in the list below.

1. Similarity of cooperating parts.
    a. In fraternal cooperation, the cooperating parts are on a fundamental level alike. Genes cooperate with genes, cells cooperate with cells, nerve cells with nerve cells, individuals with individuals, and memes with memes.
    b. In egalitarian cooperation, the parts are not alike. Different kinds of organelles cooperate in a cell, different species cooperate in symbiotic relationships, and different professions cooperate in a society.
2. On each level, there exist control mechanisms that handle constituent parts that, having lost their independence, start acting egoistically, i.e. that start functioning in a way furthering their own benefit at the detriment of the whole.
    a. In fraternal cooperation, such internal control mechanisms exist for culling or correcting overly 'selfish' genes, cancerous cells, criminal individuals, and bad ideas.
    b. In egalitarian cooperation, mutual dependence functions as a similar control mechanism—breaking cooperation entails an immediate loss.
3. On every level, there is a surplus of redundant parts, something resulting in robustness and resilience—many parts can break without causing the whole to fail.
4. The constituent cooperating parts do not always reap immediate benefits from the cooperation, but sometimes the cooperation instead benefits some other, underlying unit which is the one that really benefits from the cooperation. This

can result in cooperation that at first view can seem self-sacrificial. To solve such puzzles, it is necessary to answer the question '*For whose benefit?*' *All* cooperation benefits the copying of the replicators—genes or memes.
5. Emergent properties, where the whole is more than just the sum of its constituent parts, are a result of specialization in the constituent parts.
6. Cooperation creates order from disorder by repeating the same procedure over and over. Genes, cells and memes are copied; individuals reproduce. Genes are structured in DNA, cells are structured in organs, neurons are structured in brain components, humans are structured in groups (e.g. ethnic or professional groups), memes are structured into different fields of knowing or cultural constructs.

Synergistic cooperation is a common theme through all living beings. It has its own rules and patterns that are much alike whatever level is analyzed.

How to Harness Idea Collectives

Cooperation between ideas is not necessarily something beneficial. Like other natural phenomena there is nothing intrinsically good about the cooperation of ideas—it just happens. But if we understand cooperation between ideas better we can be more watchful about their negative effects.

Certain ideas can, for example, make their bearers deny their own sensory information for the sake of a 'cause'. This quotation from Ignatius Loyola is case in point: 'To be right in everything, we ought always to hold that the white which I see, is black, if the Hierarchical Church so decides it' (Loyola 1548). So white is black just because some authority says so? What a meme, that can coerce you to such willful blindness.

Sometimes it is even worse, sometimes we are governed by memes to such an extent that it is deadly. Cultural systems have motivated people to the ultimate self-sacrifice, to meet death for the sake of the 'cause'.

Think, for example, of Nazism, communism, nationalism, Christendom and Islam—but also of freedom of speech, democracy, justice and freedom. None of these teachings are separate ideas, but whole systems of interdependent ideas.

All these cultural systems are united in containing a core 'truth' that is larger than the individual. This can be a truth such as 'our race should prevail in the struggle between races', 'class society should be abolished', 'our nation is the greatest nation on earth', or 'this life is a qualification round for life after death.'

Humans have existed in our current form for about 200,000 years. The specific versions of religion that Christianity and Islam are have existed only for about one percent of that time. The idea of the nation as a home for all who belong together there and then, as a home worth dying for, is, according to historians, only as old as the French and American revolutions. During the last 100–200 years we have also seen the emergence of political ideas about preserving our 'race' or abolishing class society. These ideas are pathetically young in relation to humanity.

But this is also true for ideas to be cherished, such as Voltaire's viewpoint, that even though you do not share someone's viewpoint, it is right to die for their right to express it. How bizarre a thought is that—to be willing to die for *someone else's* possibility to spread their ideas? This is how strongly an idea can influence behavior. Freedom of speech, democracy, justice and freedom are also cultural systems with built-in mechanisms for their own propagation. Yet they somehow feel better than Nazism and communism, but why?

Effects of idea collectives on society are measurable. In this way they can be weighed against each other and be compared. Nazism and communism have no positive historical correlations with better health, increased well-being and increased economic welfare of all, but instead with suppression, suffering, war and death. When it comes to religion, the religiosity of a country diminishes with the level of economic and medical progress. The most affluent and well-off people live in countries with a high degree of freedom of speech, democracy, justice and freedom.

This manner of measuring the impact of ideas—the method of comparing theories with reality—is probably the most useful idea that humanity has ever come up with: *the scientific method.*

The brains of individual humans, with their impressive hardware and an updated cultural operating system installed, are capable of dreaming up the most extraordinary ideas; even more so when they are connected in networks. However, many of these ideas are just fantasies that die when meeting reality: perpetuum mobiles, horoscopes, time travel, angels and auras, to name but a few.

What determines if a theory survives or ends up on the scrap-heap of history is the meeting with reality, or the meeting with contrary theories that themselves have not been falsified through their meeting with reality. This method of systematically culling ideas has its origin in ancient Greece but gathered momentum during the scientific revolution of the 17th century.

The scientific method has led to an accumulation of a mass of knowledge that is a better description of the world around us than any earlier. This conscious selection—*domestication*—of our memes with the aid of reality is what has led to the tremendous scientific and technological advances that we have witnessed over the last couple of centuries. We may have been domesticated by our memes, but if we think about it a bit, we can cunningly reciprocate and apply the process back on them.

It isn't so much individual people who cull bad ideas from good—instead it is large human networks. No single researcher had to change her mind for Einstein's general theory of relativity to become a more general description of gravity than Newton's law of gravity. Many changed their minds, sure, but the central point is that *the scientific community* over time could see that Einstein's theory was a better explanation of reality than Newton's.

The power of using a proven better method to cull bad ideas from bad—science—makes our current society better prepared than any earlier to handle the problems and challenges that lie ahead. It is therefore not really problematic that we humans together make up the playing field on which genes competition with genes, ideas competition with ideas, and genes competition with ideas takes place.

Good ideas bring better well-being and increase our collective happiness. Political, ideological and religious ideas are tested in this manner as well. Not in consciously controlled forms, governed by a single person of an individual institution, but through the meeting of ideas and reality and the total impact the ideas have on the well-being of the nodes that are spreading them—our individual happiness.

And what higher purpose than this do we need? Long-term, sustainable well-being for as large part of the human population as possible, for us and our descendants, isn't that the best practical measure of ideas? Alternatively, we'll be outsmarted by our replicators.

> The only thing that will redeem mankind is cooperation.
>
> Bertrand Russell: Human society in ethics and politics (1954)

References

Darwin, C. (1871). *The descent of man, and selection in relation to sex* (1st ed.). London: John Murray.

Dawkins, R. (1976). *The selfish gene*. Oxford: Oxford University Press.

Godfrey-Smith, P. (2009). *Darwinian populations and natural selection*. Oxford: Oxford University Press.

Loyola, I. (1548). *Spiritual excercises*. no. 365.

Queller, D. C. (1997). Cooperators since life began. *The Quarterly Review of Biology, 72*, 184–188.

Russell, B. (1954). *Human society in ethics and politics*. Routledge.

Index

P. Lindenfors, *For Whose Benefit?*, DOI 10.1007/978-3-319-50874-0

Kolibri 360 GmbH
Ferdinand-Jühlke-Straße 7
99095 Erfurt, Deutschland
produktsicherheit@kolibri360.de